Springer Tracts in Modern Physics 75

Ergebnisse der exakten Naturwissenschaften

Manuscripts for publication should be addressed to:

G. Höhler
Institut für Theoretische Kernphysik der Universität Karlsruhe
75 Karlsruhe 1, Postfach 6380

Proofs and all correspondence concerning papers in the process of publication should be addressed to:

E. A. Niekisch
Institut für Grenzflächenforschung und Vakuumphysik
der Kernforschungsanlage Jülich, 517 Jülich, Postfach 365

R. Claus L. Merten J. Brandmüller

Light Scattering by Phonon-Polaritons

With 55 Figures

Springer-Verlag
Berlin Heidelberg New York 1975

Prof. Dr. J. Brandmüller   Dr. R. Claus

Sektion Physik der Universität München
8 München 40, Schellingstraße 4/IV

Prof. Dr. L. Merten

Fachbereich Physik der Universität Münster
44 Münster, Schloßplatz 7

ISBN 3-540-07423-6 Springer-Verlag Berlin Heidelberg New York
ISBN 0-387-07423-6 Springer-Verlag New York Heidelberg Berlin

Printing and bookbinding: Brühlsche Universitätsdruckerei, Gießen

Preface

Our understanding of the properties of phonon polaritons has reached a level that allows numerous applications in the field of chemical analysis of the experimental methods developed. The present volume is intended as an introduction to the field and is written primarily for experimentalists. Theories have normally been included only to the extent they have been verified. Representative experiments demonstrating the different theoretically derived effects are described in detail.

The book begins with a brief review of the Raman effect and its applications, including some historical remarks. Grouptheoretical aspects of light scattering by phonons are summarized in Chapter 2. Chapter 3 gives an elementary description of the simplest model for phonon polaritons in ideal crystals, starting from the Born - von Kármán model which is assumed to be known from any textbook on solid-state physics. Chapter 4 deals with the theory and experimental methods for polyatomic crystals of arbitrary symmetry. Finally, some related subjects, such as stimulated scattering and surface effects, are treated in Chapter 5.

Various parts of the text have been written by the authors Brandmüller (B), Claus (C), and Merten (M) as follows: Sections 1.1 to 2.6 (B), 3.1 to 4.9 (C), 4.10 to 4.14 (M), 5.1 to 5.5 (C), Appendix 1 (M), Appendices 2 and 3 (C), and Appendix 4 (B/C). We want to thank G. Borstel, J. Falge, H. W. Schrötter, and F.X. Winter for stimulating discussions. We thank W. Kress, Stuttgart for references and J.F. Scott, Boulder, Colorado for critical remarks.

Munich, December 1974

R. Claus
L. Merten
J. Brandmüller

Table of Contents

1. Introduction

1.1 WHAT ARE POLARITONS?

Because ideal crystalline materials show translational symmetries the eigenfunctions of excited states in such materials can be represented by plane waves according to Bloch's theorem /1,5/. The excited states frequently are associated with an electric (or magnetic) polarization field so that polarization waves will be generated. The polarization waves on the other hand couple with electromagnetic waves described by Maxwell's equations. Coupled excited states of this type have become known as polaritons. The most important types of polarization waves in crystals are exciton waves /3,6/, plasma waves /4/, and long-wavelength infrared-active optical lattice waves. 'Long-wavelength' means that the wavelength λ is very large compared with the short-range interatomic forces, see 4.13.

The term 'polariton', however, also stands for the energy quanta of the excitations in question. Thus a polariton is a quasi-particle consisting of a photon coupled with an exciton, plasmon or a long-wavelength polar optical phonon, see 4.14. The different types of polaritons are distinguished by the terms exciton-like, plasmon-like or phonon-like.If one includes the coupling via a magnetic field, the coupled state magnon-photon can also be regarded as a special type of polariton, see 3.4. Although magnon-polaritons have been predicted theoretically /5/, the existence of these quasi-particles has not been established experimentally so far.

In recent days there has been a trend to denote any coupled states between photons and elementary excitations in matter as polaritons /300/.

Some attempts have been made in order to develop a general theory so that the different types of polaritons cited above are obtained as special cases, see for instance the article by Hopfield /2/. A final form of such a theory, however, still is missing.

We shall restrict our discussion in this volume to phonon-polaritons. For simplicity, we are thus frequently going to use the term 'polariton' only in this specialized sense. The existence of mixed excited states which are partly mechanical and partly electromagnetic has first been predicted by Huang /36/ in 1951. The denomination 'polariton', however, was not introduced until 1958 /2/. We also refer to the nice article 'Who Named the -ON's?' by Walker and Slack /292/. The authors claim that Fano was responsible for the 'polariton-concept'. Fano, however, only presents a first quantum mechanical treatment of coupled electromagnetic and lattice fields in 1956 /302/.

Additional Literature

We refer to review articles by Loudon /31, 160, 314/, Hopfield /3, 303/, Pick /80, 310/, Scott /71/, Merten /74, 78/, Claus /75, 76/, Brandmüller and Schrötter /35/, Mills and Burstein /300/, Barker and Loudon /73/, Born and Huang /62/, Lax and Nelson /311/, Burstein /313/. The Proceedings of the first Taormina Research Conference on the Structure of Matter: Polaritons /316/ are mentioned here, too.

For those who desire a quick introduction to the field, we refer to an article by Claus: 'Dispersion Effects of Polar Optical Modes in Perfect Crystals' /77/. This article represents a comprehensive form of the following text.

1.2 THE RAMAN EFFECT

The existence of polaritons was first demonstrated experimentally by Henry and Hopfield in 1965 on GaP /7/ by means of the Raman effect. Since then the Raman effect has remained the most important experimental method giving information on the physics of these quasi-particles. We therefore present a short review including some historical remarks. Other experimental methods have been described in 5.3 and 5.4 and, for instance, in /59, 283/.

Frequency-shifted scattered radiation from matter was first reported by Raman /8/ and somewhat later by Landsberg and Mandelstam /347/ in 1928. Raman observed that the light from a mercury lamp was scattered in such a way by liquid benzene that the spectrum contained more lines than that of the mercury lamp itself.

An elementary Stokes scattering process takes place when an incident photon with the energy $\hbar\omega_i$ is annihilated and another photon $\hbar\omega_s$ is created simultaneously with a quantum $\hbar\omega$. With Anti-Stokes scattering on the contrary a, for instance, thermally excited quantum $\hbar\omega$ is annihilated. Energy conservation requires

$$\hbar\omega_i = \hbar\omega_s + \hbar\omega \qquad \text{for Stokes processes}$$

and

$$\hbar\omega_i = \hbar\omega_s - \hbar\omega \qquad \text{for Anti-Stokes processes.} \tag{1-1}$$

The energy difference between the incident and scattered photons corresponds to a change of the energy state of the material caused by the interaction with light. In 1923 a relation

$$\hbar\omega_i + E_k = E_n + \hbar\omega_s \tag{1-2}$$

was derived by Smekal /9/ from quantum-theoretical considerations. Herein E_k and E_n stand for the energy eigenvalues of a medium before and after a light-scattering process, respectively. Smekal thus predicted the existence of frequency-shifted scattered radiation.

In a classical model the power of radiation emitted by a dipole is

$$I = (2/3c^3)\langle|\ddot{\underline{P}}|^2\rangle \quad , \tag{1-3}$$

where $\underline{P}$ denotes the oscillating electrical dipole moment and <> means time-averaged. When regarding a molecule with the polarizability α the dipole moment induced by the electric field $\underline{E}$ of the incident radiation becomes in a linear approximation

$$\underline{P} = \alpha\,\underline{E} \quad . \tag{1-4}$$

α is a second rank tensor. The approximation is justified for 'small' field strengths $\underline{E}$, i.e. for spontaneous Raman scattering, see 5.1. For molecular vibrations or rotations the components of α are periodical functions in time with fixed eigenfrequencies modulating the incident electromagnetic wave. This is the classical interpretation of the Raman effect. In practice it is convenient to use linearly polarized incident light (lasers). The electric field vector $\underline{E}$ then can be arranged to have one nonvanishing component only so that (1-4) becomes explicitly, for instance

$$\begin{pmatrix} P_x \\ P_y \\ P_z \end{pmatrix} = \begin{pmatrix} \alpha_{xx} & \alpha_{xy} & \alpha_{xz} \\ \alpha_{yx} & \alpha_{yy} & \alpha_{yz} \\ \alpha_{zx} & \alpha_{zy} & \alpha_{zz} \end{pmatrix} \begin{pmatrix} 0 \\ 0 \\ E_z \end{pmatrix} = \begin{pmatrix} \alpha_{xz}\, E_z \\ \alpha_{yz}\, E_z \\ \alpha_{zz}\, E_z \end{pmatrix} \quad . \tag{1-5}$$

Scattered radiation on the other hand is analyzed advantageously with a polarizer before the entrance slit of the spectrometer. This makes it possible to record the light scattered according to one certain tensor element of α, for instance, α_{zz} separately. From (1-3) follows

$$I \sim \omega_i^4 (\alpha_{zz})^2 \quad . \tag{1-6}$$

If $\underline{e}_i$ denotes a unit vector parallel to the polarization of the incident laser radiation and $\underline{e}_s$ correspondingly describes the polarization of the observed scattered light for a constant laser frequency ω_i (1-6) can be written alternatively

$$I \sim |\underline{e}_i \; \alpha(\omega) \; \underline{e}_s|^2 \quad , \tag{1-7}$$

$\alpha(\omega)$ is the frequency dependent polarizability tensor from (1-4). Eq.(1-7) describes the total scattered radiation including Rayleigh, Raman and higher-order Raman scattering.

In a crystalline medium the molecular polarizability α must be replaced by the susceptibility χ. χ is a macroscopic property of the material which will be discussed in more detail when considering the crystal symmetry, see 2.3 to 2.6. The scattering intensity in crystals becomes

$$I \sim |\underline{e}_i \, \chi(\omega) \, \underline{e}_s|^2 \quad . \tag{1-8}$$

For further discussion of scattering intensities in crystals,we refer to 4.12 and Appendix 4.

Kramers and Heisenberg /10/ derived a quantum-theoretical relation for the scattering intensity by transferring the classical model via the correspondence principle. This procedure can be summarized in the following way: The radiation is treated classically whereas the medium is treated quantum-mechanically. The change of the eigenfunctions of the material caused by the exciting light wave is derived by perturbation theory. This determines the electric dipole moment. In order to calculate the scattering intensity, (1-3) has hypothetically to be introduced. Dirac /11/ in 1927 eliminated this inconsistency by treating also the radiation field quantum-mechanically. In both models the scattering intensity is determined by

$$I_{kn} = (64\pi^4/3c^3)(\nu_i + \nu_{kn})^4 |\underline{C}_{kn}|^2 \tag{1-9}$$

where

$$\underline{C}_{kn} = (1/h) \sum_r \left[\frac{(\underline{A}\langle k|\underline{P}|r\rangle)\langle r|\underline{P}|n\rangle}{\nu_{rk} - \nu_i} + \frac{\langle k|\underline{P}|r\rangle(\underline{A}\langle r|\underline{P}|n\rangle)}{\nu_{rn} + \nu_i} \right] \quad .$$

$\underline{A}$ and ν_i stand for the amplitude and frequency of the incident radiation, respectively. Two photons are involved in a linear scattering process. ν_{rk} and ν_{rn} are the frequency differences of the levels indicated by the subscripts r, k, n, respectively. The quantities $\underline{A}\langle k|\underline{P}|r\rangle$ and $\underline{A}\langle r|\underline{P}|n\rangle$ are scalar products. The medium is changed from the energy level k to n,and ν_{kn} accordingly denotes the frequency shift of the scattered radiation,which is identical with the frequency of the Raman line observed, see (1-1). The sum has to be taken over all energy levels of the system; they play an important role as intermediate states of the scattering process. The energy of the incident photons must not be so large as to reach one of the levels r. The Raman effect (except for full resonance, see 1.4 and /155/) is not a sequence of absorptions and emissions as is, for instance, fluorescence.

The main difficulty for numerical calculations on the basis of (1-9) is that the complete set of energy levels in general is not known even for the most simple systems. Placzek /12/ therefore developed a semi-classical model for the vibrational Raman effect which has become known as the 'polarizability theory'.

A molecule with fixed nuclei is supposed to have an electronic polarizability α_o. If the nuclei are in motion, because of molecular vibrations, the polarizability is modified. Placzek could show that α is only a function of the positions of the nuclei but not of their velocities if

1) $\nu_e - \nu_i \gg |\nu_{nk}|$, i.e. the frequency of the incident radiation ν_i has to be far enough away from any electronic transition frequencies ν_e. $\nu_e - \nu_i$ must be large compared with the splittings and shifts of the electronic ground state due to the vibrations of the nuclei.
2) $\nu_i \gg |\nu_{nk}|$, i.e. the exciting frequency ν_i must be large compared with the vibrational frequencies in the electronic ground state.
3) The electronic ground state may not be essentially degenerate.

Placzek succeeded in finding a relation for the scattering intensity containing only quantities referring to the electronic ground state. Any influence of other states is described by the polarizability α and its dependence on the nuclear coordinates. α can be expanded into a series in terms of the normal coordinates Q_j as follows

$$\alpha(\underline{Q}) = \alpha_o + \sum_j \left(\frac{\partial\alpha}{\partial Q_j}\right)_o Q_j + \frac{1}{2} \sum_{j,k} \left(\frac{\partial^2\alpha}{\partial Q_j \partial Q_k}\right)_o Q_j Q_k + \ldots \qquad (1\text{-}10)$$

The Raman scattering intensity of a molecule can be expressed by means of the trace a_j and the anisotropy γ_j^2 of the tensor containing the first derivatives of the polarizability with respect to the normal coordinates Q_j (the linear term in (1-10)). If the scattered radiation is observed in a direction perpendicular to the incident light and furthermore the exciting light is polarized perpendicular to the direction of observation, the following equation holds:

$$I_{\parallel}(\nu_i \mp \nu_j) = \frac{2^4 \pi^3 h (\nu_i \mp \nu_j)^4}{3^3 \; 5c^4 \nu_j} g_j \cdot I_o \left[\underline{+}1 \mp \exp\left(\mp \frac{h\nu_j}{kT}\right)\right]^{-1} (5a_j^2 + 4\gamma_j^2) \quad . \qquad (1\text{-}11)$$

If, on the contrary, the exciting light is polarized parallel to this direction the corresponding relation $I_{\perp}(\nu_i \mp \nu_j) = \ldots$ is obtained by simply exchanging the last bracket with $3\gamma_j^2$. The experiment in the latter case has to be carried out with an analyzer parallel to the polarization of the incident light. The upper signs hold for Stokes processes and the lower ones for Anti-Stokes processes. ν_j is the eigenfrequency of the normal coordinate Q_j and g_j the corresponding degree of degeneration. I_o stands for the power of the incident radiation per cm^2 at the position of the molecule. Because it is still very difficult to determine all the tensor elements $(\partial\alpha/\partial Q_j)\cdot Q_j$ explicitly for free molecules, important information is usually derived from the depolarization ratio ρ_s, which for linearly polarized exciting light (lasers) is defined

$$\rho_s = I_{\perp}/I_{\parallel} = 3\gamma_j^2/(5a_j^2+4\gamma_j^2) \quad . \tag{1-12}$$

It is an important task of molecular spectroscopy to assign the observed normal vibrational modes to the different symmetry species of the molecular point groups. Only for totally symmetric vibrations does the trace a_j not vanish. Accordingly the depolarization ratio becomes $0 \leq \rho_s < 0.75$. $\rho_s = 0$ holds for molecules with cubic symmetry. For all non totally symmetric species the trace vanishes so that the depolarization ratio becomes $\rho_s = 0.75$.

The number of normal vibrations of every species can be calculated by group-theoretical methods /13, 14/. Correspondingly, selection rules for Raman scattering and IR absorption can be derived for the fundamentals as well as for harmonics and combinations, see 2.5.

1.3 DIFFERENT CAUSES OF THE RAMAN EFFECT

There are quite a lot of different excitations which according to (1-2) can cause a Raman effect. E_k and E_n may stand for energy eigenvalues of a <u>rotator.</u> In linear Raman scattering the condition for the appearance of rotational lines on both sides of the unshifted Rayleigh-line is a finite optical anisotropy. From rotational spectra of molecules information is obtained on the moments of inertia and thus the interatomic distances. Information can also be obtained on the nuclear-spin, the spin-spin, and spin-rotational interactions e.g. for O_2 /15/.

The internal motions of N atoms in a molecule can be described by 3N-6 independent harmonic oscillators by means of normal coordinates. For linear molecules there are only 3N-5 normal coordinates. The energy eigenvalues in (1-2) then determine vibrational Raman lines. In (1-10) the sums over j and k have to be taken from 1 to (3N-6) or from 1 to (3N-5), respectively. The symmetry properties of normal coordinates essentially depend on the structure of the molecule as described by the point group. Thus it is possible to get information on molecular structures from Raman and IR spectra. The selection rules are of great importance. The appearance of a vibrational mode in the linear Raman effect is determined by the symmetry properties of the polarizability tensor whereas the selection rules for IR absorption are determined by the symmetry properties of the electric dipole moment which is a polar vector. As a result, for molecules with an inversion center, normal vibrations appearing in the linear Raman spectrum are forbidden in the infrared and vice versa, see 2.5. All vibrational Raman lines of free molecules have a rotational structure superimposed. The only exceptions are the totally symmetric vibrations of molecules with cubic symmetries. The rotational structures in general can only be resolved in gases. Prevented rotations (librations) in liquids and solids are responsible for the different profiles of Raman lines.

E_k and E_n in (1-2) may also denote different electronic energy levels. This was initially verified experimentally by Rasetti in 1930 on NO /16/. Welsh et al. /17/ later could resolve the rotational structure of the electronic Raman line in this material in detail appearing at $\sim 125\ cm^{-1}$. Light scattering by electronic levels has become of great importance for investigations of solids. Here normally it is referred to as light scattering by excitons /18/.

Information on lattice dynamics of single crystals is obtained by means of light scattering from transverse (TO) and longitudinal (LO) optical phonons; see 2.1, 2.2 and the detailed discussions in Chapters 3 and 4.

In semiconductors there are longitudinal waves of the free electron plasma. The corresponding quantized excited states are referred to as plasmons, which consequently may cause Raman scattering too /19 - 21, 9/. When a magnetic field is applied perpendicular to their wave vectors, plasmons can couple with transverse polar modes. This happens because the Lorentz force induces a quasi-transverse character of the plasmons. Coupled states of this type have become known as plasmaritons. For detailed discussions see 5.2.

Ferro-and antiferromagnetic properties of matter are determined by the existence and arrangement of magnetic dipole moments caused by the electron spins. The energy quanta of spin waves have become known as magnons, which according to (1-2) again give rise to Raman scattering, see for instance /22/.

If, finally, a magnetic field is applied to a conductor or semiconductor, translations of the free electrons are superimposed by rotations at cyclotron frequencies. The corresponding quantized energy levels are referred to as Landau levels. Inelastic scattering of light again causes a Raman effect which has been experimentally verified in good agreement with the theory in InSb /23/.

Additional Literature

Porto, S.P.S.: Light Scattering with Laser Sources /315/.
Anderson, A.: The Raman Effect /286/.
Koningstein, J.A.: Introduction to the Theory of the Raman Effect /333/.
Poppinger, M.: Magnonen, Phononen und Excitonen von MnF_2 /334/.
Szymanski, H.A.: Raman Spectroscopy, Theory and Practice /335/.
Woodward, L.A.: Introduction to the Theory of Molecular Vibrations and Vibrational Spectroscopy /336/.

1.4 RESONANCE RAMAN EFFECT

As has been pointed out in 1.2, Placzek's polarizability theory holds only when the exciting line is not located too close to an electronic absorption band. When this does happen the full quantum-mechanical treatment has to be applied. The first term in the bracket of (1-9) becomes large if the exciting frequency ν_i becomes close to ν_{rk}. In the case of resonance, the scattering intensity thus will be very strong. It should be noted, however, that the scattering cross-sections of different Raman lines do not increase in the same way. It turns out that only some of them (mainly the totally symmetric) will be intense. Theories for the resonance-Raman effect have been developed by, for instance, Schorygin /24/ and Behringer /25, 305/. Resonance-Raman scattering is of great importance for studies of heavily absorbing targets.

Additional Literature

Bendow, B., Birman, J.L.: Polariton Theory of Resonance Raman Scattering in Insulating Crystals /220/.

Kiefer, W.: Laser-excited Resonance Raman Spectra of Small Molecules and Ions /337/.

Behringer, J.: Experimental Resonance Raman Spectroscopy /338/.

1.5 NONLINEAR RAMAN EFFECTS

When irradiating a crystal with a giant pulse laser very high electric field strengths at optical frequencies can be achieved in the material. The electric polarization then cannot be considered as a linear function of the field strength as in (1-4). Terms of higher order in $\underline{E}$ must in addition be taken into account, see 5.1. Corresponding phenomena are referred to as nonlinear optics. Nowadays, in Raman spectroscopy we distinguish three different nonlinear effects:

a) The stimulated Raman effect is of greatest importance for our subject. In the early days Kerr cells containing nitrobenzene were used in order to construct giant pulse lasers. In the spectrum of the radiation from such lasers a strong line shifted by 1345 cm^{-1} away from the ruby line at 694.3 nm was detected in 1962 /16/. G. Eckhardt identified this line as originating from the totally symmetric NO_2 valence vibration of nitrobenzene /27/. Only this line, being the strongest one in the linear Raman spectrum of NO_2, was observed. The effect has become known as 'stimulated Raman effect' and has been observed in many materials and all states of matter since then. The most characteristic differences to the linear Raman effect are: only 1, 2 or at most 3 lines of the linear Raman spectrum (which may consist of a very large number of lines) are generated. The scattering intensity of these lines is of the order of the Rayleigh-line intensity. Several harmonics of the stimulated lines are frequently observed, see 5.1. Stimulated Anti-

Stokes lines and their harmonics are generated with similar intensities. The elementary scattering process involves four photons as described by the wave vector relation $2\underline{k}_i = \underline{k}_S + \underline{k}_A$, see (5-17) and (5-21) and their derivation. The intensity threshold observed for the generation of stimulated Raman scattering is essentially determined by the optical Kerr constant which is responsible for the self-focusing of the laser beam.

Stimulated Raman scattering has become of great importance for the construction of 'Raman lasers'. Intense coherent radiation at different wavelengths may be generated by using different materials. When generating stimulated radiation from polaritons, one obtains a 'polariton-laser' tunable over a certain frequency region, see 5.1.

The remaining nonlinear Raman effects have not yet become of special interest for polaritons.

b) The inverse Raman effect was also detected in liquids /278/. When a medium is irradiated simultaneously by intense monochromatic light from a giant pulse laser and by a continuum, sharp absorption lines are observed on the anti-Stokes side of the laser line, and under special conditions also on the Stokes side /279/. Gadow et al. /330/ have recently studied the inverse Raman effect on single crystalline $LiIO_3$ and $LiNbO_3$, see also Kneipp et al. /353/.

c) The hyper-Raman effect appears when the electric field of the exciting radiation is very strong. Higher-order terms of the induced dipole moment again become significant /280/, see (1-4) and 5.1

$$P_\rho = \sum_\sigma \alpha_{\rho\sigma} E_\sigma + \frac{1}{2} \sum_{\sigma,\tau} \beta_{\rho\sigma\tau} E_\tau E_\sigma + \frac{1}{6} \sum_{\sigma,\tau,\upsilon} \gamma_{\rho\sigma\tau\upsilon} E_\upsilon E_\tau E_\sigma + \ldots \quad . \qquad (1\text{-}13)$$

The 'first and second hyperpolarizabilities' $\beta_{\rho\sigma\tau}$ and $\gamma_{\rho\sigma\tau\upsilon}$, respectively, lead to second and third harmonic light scattering at the frequencies $2\omega_i$ and $3\omega_i$ (hyper-Rayleigh scattering). Their derivatives with respect to the normal coordinates correspondingly lead to nonlinear inelastic light scattering at $2\omega_i \pm \omega$ and $3\omega_i \pm \omega$ (hyper-Raman scattering). ω_i and ω denote the frequencies of the incident laser light and an elementary excitation in the material, respectively.

In addition to these nonlinear Raman effects,the generation of

Stokes and anti-Stokes radiation by mode-mixing in nonlinear materials has become of increasing interest (e.g. 'coherent Anti-Stokes Raman spectroscopy'). For discussions of these phenomena we refer to 5.1.

We finally cite review articles on the Raman effect in crystals: /21, 29 - 35, 300/.

2. Raman Scattering by Optical Phonons

2.1 THE PHONON WAVE-VECTOR

The wave-vector $\underline{k}$ propagates in the direction of the wave normal. The absolute value of $\underline{k}$ for photons in vacuum is

$$|\underline{k}| = k = \omega/c \quad , \tag{2-1}$$

where $\omega = 2\pi\nu$ is the radian frequency of the electromagnetic wave and c the velocity of light in vacuum. If the vacuum wavelength is called λ, trivially

$$c = \lambda\nu \tag{2-2}$$

and the wave number is

$$k = 2\pi/\lambda \quad . \tag{2-3}$$

In a dielectric medium of refractive index n the vacuum velocity of light is replaced by the phase velocity c/n. Thus (2-1) is replaced by

$$k = n\omega/c \tag{2-4}$$

and (2-2) by

$$c/n = \lambda\nu \quad . \tag{2-5}$$

If, finally, we denote the wavelength in the dielectric medium by λ_n, the wave number can generally be written

$$k = 2\pi/\lambda_n = 2\pi n/\lambda \quad . \tag{2-6}$$

Note that according to (2-4) the refractive index can always be written in the form

$$n = ck/\omega \quad . \tag{2-4a}$$

This identity is of great importance in all the dispersion relations to be discussed below.

The wave vector of a photon is directly correlated to its momentum. The magnitude of the momentum is $p = mc$ and the photon energy $\hbar\omega = mc^2$. By eliminating m, we obtain $p = \hbar k$. The photon momentum is defined as a vector in the direction of the wave vector

$$\underline{p} = \hbar\underline{k} \quad . \tag{2-7}$$

Because the propagation of phonons is not associated with mass transport, phonons do not have a real momentum. Therefore, the quantity $\hbar\underline{k}$ for phonons has become known as 'quasi-momentum' /37/. Süssmann has discussed this quasi-momentum in detail /38/. From a microscopic point of view, inelastic scattering of a photon by a phonon is associated with a local deformation of the lattice, in other words an atom becomes displaced from its equilibrium position in a certain direction determined by a displacement vector $\underline{u}$. $\underline{u}$ is small compared with the lattice constant in a linear scattering process. The deformation is transmitted through the medium. This process can be described by plane waves with wave-vectors perpendicular to $\underline{u}$. In fact, for infinite phonon wavelengths the quasi-momentum $\hbar k$ is the macroscopic momentum of the crystal. Inelastic photon-phonon scattering is therefore described by a wave-vector relation, often referred to simply as 'momentum conservation'

$$\hbar\underline{k}_i = \hbar\underline{k}_s \pm \hbar\underline{k} \quad , \tag{2-8a}$$

$$\underline{k}_i = \underline{k}_s \pm \underline{k} \quad . \tag{2-8b}$$

$\underline{k}_i$, $\underline{k}_s$ and $\underline{k}$ are the wave-vectors of the incident and scattered photons, and the phonon, respectively. The upper signs in (2-8) hold for a Stokes process and the lower ones for an Anti-Stokes process. Energy conservation correspondingly requires

$$\omega_i = \omega_s \pm \omega \qquad \text{(see (1-1))}. \tag{2-9}$$

2.2 OBSERVATION OF OPTICAL PHONONS BY RAMAN SCATTERING

We are going to determine the magnitude of wave-vectors for phonons involved in Raman scattering processes of first order.

When using an exciting (laser) wavelength in the visible region of the order $\lambda_i = 500$ nm $= 5 \times 10^{-5}$ cm, the magnitude of the corresponding photon wave vector is $k_i = 2\pi n/\lambda_i \approx 10^5$ cm^{-1}, provided the refractive index of the material is not too different from 1. The relative wave numbers of phonons observed by Raman spectroscopy in general are in the range 0 to 4000 cm^{-1}. From (2-9) it can be seen that ω_s therefore remains of the same order of magnitude as ω_i and $k_s \approx k_i$. In a right angle scattering process with $\measuredangle (\underline{k}_i, \underline{k}_s) = \pi/2$ the phonon wave vector becomes $\sqrt{2} \cdot k_i$ due to (2-8). This is again of the order $k \approx 10^5$ cm^{-1}. Phonon wave-vectors at the boundary of the first Brillouinzone (1BZ), on the other hand have a (maximum) magnitude $k = \pi/2a$. Here 2a denotes the lattice constant, which is normally ≈ 0.1 nm. Zone-boundary phonon wave-vectors are therefore of the order of $k \approx 10^8$ cm^{-1}. The wave-vectors of (fundamental) phonons observed by Raman scattering are about 3 orders of magnitude smaller and thus located almost in the center of the 1BZ ($k \approx 0$).

2.3 FACTOR GROUPS ANALYSIS

Conventional factor goups analysis (FGA) holds for $k \equiv 0$. As can be seen from 2.2, FGA is a good approximation for nonpolar modes observed by first order Raman scattering. When the phonon spectrum of a crystal is to be examined, FGA provides information on the number of modes expected for the different symmetry species. In the spectra of polar modes the degeneracies predicted by FGA for $k = 0$ are normally removed. Taking into account TO-LO splittings and directional dispersion, however, which are discussed later, corresponding information concerning these modes can equally well be obtained from FGA. We need to know the structure of the elementary cell and the character table of the factor group, see 2.4. We omit detailed discussions because the corresponding methods have recently been described and illustrated by Behringer /39/.

We therefore include only a few basic remarks concerning the nomenclature of the character tables.

Additional Literature

Zak, J., Casher, A., Glück, M., Gur, Y.: The Irreducible Representation of Space Groups /339/.
Behringer, J.: Raman-Spektren von Kristallen (k-abhängige Gruppentheorie) /340/.
Miller, S.C., Love, W.F.: Tables of Irreducible Representations of Space Groups and Co-Representations of Magnetic Space Groups /341/.

2.4 CHARACTER TABLES OF THE POINT GROUPS

The factor group P' is isomorphous to the direction group P (or crystal-class point group), i.e. the characters of the irreducible representations for both groups are identical. The point groups are well known from molecular spectroscopy, see for instance /13, 14/. The symmetries of non-linear molecules can be described by the 43 point groups listed in Table 1. Only 32 of these (indicated by a black circle) are of importance in crystallography. They contain symmetry elements with n-fold rotational or rotational-mirror axes, where n is only = 1, 2, 3, 4, 6. The 32 point groups cause the classification of crystal structures into 32 crystal classes, see Table 5 in /39/. The international symbols for the point groups introduced by Hermann and Mauguin have been added. The former system of nomenclature was used in the International Tables for X-Ray Crystallography, Vol. I, 3rd ed. 1969 /40/. Because of the large number of group symbols it is convenient to denote the order of both point and space groups by the group symbol in square brackets: ($[P]$, $[D_3]$). The real subgroups of the point groups are listed in column 6 of Table 1.

Symmetry operations cannot easily be illustrated graphically. Therefore the concept of symmetry elements has been introduced. Some, but not all aspects of symmetry operations are described in this way.

Ordning $g=[P]$	Abstract group G	Pointgroup P Schoenflies	Pointgroup P International	Real subgroups of P	Nr.
1	$G_1=C_1$	• C_1	1	–	1
2	$G_2=C_2$	• C_2	2	–	2
		• $C_i=S_2$	$\bar{1}$	–	3
		• $C_s=S_1=C_{1h}$	$m=\bar{2}$	–	4
3	$G_3=C_3$	• C_3	3	–	5
4	$G_4^1=C_4$	• C_4	4	C_2	6
		• S_4	$\bar{4}$	C_2	7
	$G_4^2=D_2=V$	• $D_2=V$	222	C_2	8
		• C_{2v}	mm2 = 2mm = mm	C_2, C_s	9
		• C_{2h}	2/m	C_i, C_2, C_s	10
5	$G_5=C_5$	C_5	5	–	11
6	$G_6^1=C_6$	• C_6	6	C_2, C_3	12
		• $S_6=C_{3i}$	$\bar{3}$	C_i, C_3	13
		• $S_3=C_{3h}$	$\bar{6}$	C_s, C_3	14
	$G_6^2=D_3$	• D_3	32	C_2, C_3	15
		• C_{3v}	3m	C_s, C_3	16
8	$G_8^1=C_8$	S_8	$\bar{8}$	C_2, C_4	17
	G_8^3	• C_{4h}	4/m	$C_i, C_2, C_s, C_{2h}, C_4, S_4$	18
	$G_8^4=D_4$	• D_4	422 = 42	C_2, D_2, C_4	19
		• C_{4v}	4mm = 4m	C_2, C_s, C_{2v}, C_4	20
		• $D_{2d}=V_d$	$\bar{4}2m$	$C_2, C_s, D_2, C_{2v}, S_4$	21
	G_8^5	• $D_{2h}=V_h$	$mmm=\frac{2}{m}\frac{2}{m}\frac{2}{m}$	$C_i, C_2, C_s, D_2, C_{2v}, C_{2h}$	22
10	$G_{10}^1=C_{10}$	$C_{5h}=S_5$	$\overline{10}$	C_s, C_5	23
	$G_{10}^2=D_5$	D_5	52	C_2, C_5	24
		C_{5v}	5m	C_2, C_5	25
12	G_{12}^2	• C_{6h}	6/m	$C_i, C_2, C_s, C_{2h}, C_3, S_6, C_6, S_3$	26
	$G_{12}^4=D_6$	• D_6	622 = 62	C_2, D_2, C_3, D_3, C_6	27
		• C_{6v}	6mm = 6m	$C_2, C_s, C_{2v}, C_3, C_{3v}, C_6$	28
		• D_{3h}	$\bar{6}m2=\bar{6}2m$	$C_2, C_s, C_{2v}, C_3, D_3, C_{3v}, S_3$	29
		• D_{3d}	$\bar{3}m=\bar{3}\frac{2}{m}$	$C_i, C_2, C_s, C_{2h}, C_3, S_6, D_3, C_{3v}$	30
	$G_{12}^5=A_4=T$	• T	23	C_2, D_2, C_3	31
16	G_{16}'	• D_{4h}	$4/mmm=\frac{4}{m}\frac{2}{m}\frac{2}{m}$	$C_i, C_2, C_s, C_{2h}, D_2, C_{2v}, D_{2h}, C_4, S_4, C_{4h}, D_4, C_{4v}, D_{2d}$	32
	$G_{16}''=D_8$	D_{4d}	$\bar{8}2m$	$C_2, C_s, C_4, D_2, C_{2v}, S_8, D_4, C_{4v}$	33
20	G_{20}'	D_{5h}	$\overline{10}m2=\overline{10}2m$	$C_s, C_2, C_{2v}, C_5, C_{5h}, D_5, C_{5v}$	34
		D_{5d}	$\bar{5}m=\bar{5}\frac{2}{m}$	$C_i, C_2, C_s, C_{2h}, C_5, S_{10}, D_5, C_{5v}$	35
24	G_{24}'	• T_h	$m3=\frac{2}{m}\bar{3}$	$C_i, C_2, C_s, C_{2h}, D_2, C_{2v}, D_{2h}, C_3, S_6, T$	36
	G_{24}''	• D_{6h}	$6/mmm=\frac{6}{m}\frac{2}{m}\frac{2}{m}$	$C_i, C_2, C_s, C_{2h}, D_2, C_{2v}, D_{2h}, C_3, S_6, D_3, C_{3v}, D_{3d}, C_6, S_3, C_{6h}, D_6, C_{6v}, D_{3h}$	37
	$G_{24}'''=D_{12}$	D_{6d}	$\overline{12}2m$	$C_2, C_s, C_3, S_4, D_2, C_{2v}, C_6, D_3, C_{3v}, D_{2d}, S_{12}, D_6, C_{6v}$	38
	$G_{24}''''=S_4=O$	• O	432 = 43	$C_2, D_2, C_4, D_4, C_3, D_3, T$	39
		• T_d	$\bar{4}3m$	$C_2, C_s, D_2, C_{2v}, S_4, D_{2d}, C_3, C_{3v}, T$	40
48	G_{48}'	• O_h	$m3m=\frac{4}{m}\bar{3}\frac{2}{m}$	$C_i, C_2, C_s, C_{2h}, D_2, C_{2v}, D_{2h}, C_4, S_4, C_{4h}, D_4, C_{4v}, D_{2d}, D_{4h}, C_3, S_6, D_3, C_{3v}, D_{3d}, T, T_h, O, T_d$	41
60	$G_{60}'=A_5=I$	I	532	$C_2, C_3, D_2, C_5, D_3, D_5, T$	42
120	G_{120}'	I_h	53m	$C_i, C_2, C_s, C_3, D_2, C_{2v}, C_{2h}, C_5, S_6, D_3, C_{3v}, D_{2h}, S_{10}, D_5, C_{5v}, T, D_{3d}, D_{5d}, T_h, I$	43

Tab.1

The 43 point groups which are of importance in molecular and crystal spectroscopy, from /340/.

The symmetry operation C_n, for instance, stands for all rotations C_n^j where $j = 1,\ldots,n-1$. The symmetry element on the other hand primarily describes the geometric structure of the operation ('center', 'axis', 'plane'). Nongeometric attributes are added to indicate the connection with the operations ('inversion', 'rotational', 'mirror', 'fourfold' etc.). The point group C_6, as illustrated by the material $LiIO_3$ /41, 42/, for instance, is cyclic. We shall make use of this point group to explain the character tables. All group elements are obtained as integer powers of only one generating element, as can be seen from Table 2. The identity is $E = C_6^6$. Every group has as many irreducible representations as there are classes of group elements. The term 'symmetry species' is used as a synonym for the mathematical term 'irreducible representation' in spectroscopy /52/. The symmetry-species symbols for point group C_6 are listed in the second column of Table 2. Detailed definitions of the symbols used for different point groups are given in /51/. The general meaning is:

A symmetry with respect to the most-fold axis,

B antisymmetry with respect to the most-fold axis,

E twofold degeneracy,

F threefold degeneracy,

G fourfold degeneracy etc.

The degree of degeneracy can be seen from the column for the identity E in the character tables. Symmetry or antisymmetry, for instance, with respect to an inversion center i is indicated by subscripts 'g' or 'u' on the symmetry-species symbol. The former are found to be only Raman-active and the latter only IR-active, (see 2.5). This is of great importance because polaritons cannot be observed directly by Raman scattering in materials with an inversion center (e.g. all alkalihalides).

The characters of the irreducible representations of the species are normally listed as shown in Table 2. The A modes are totally symmetric with respect to all symmetry elements (characters = + 1). The B modes on the other hand are symmetric with respect to E, C_3 and C_3^2 and antisymmetric with respect to C_6, C_2 and $\overline{C}_6$ (characters = - 1). In general, however, the characters are complex and cannot be interpreted intuitively, see e.g. the E modes in the table.

1	2	3						4	5	6
γ	C_6	E	C_6	$C_6^2=C_3$	$C_6^3=C_2$	$C_6^4=C_3^2$	$C_6^5=\overline{C}_6$	μ_ρ, Q^T_ρ	$\alpha_{\rho\sigma}$	Q^R_ρ
	6	1	6	6^2	6^3	6^4	6^5			
1	A	1	1	1	1	1	1	$\begin{pmatrix}0\\0\\z\end{pmatrix}$	$\frac{1}{2}\begin{pmatrix}xx+yy & xy-yx & 0\\-(xy-yx) & xx+yy & 0\\0 & 0 & 2zz\end{pmatrix}$	$\begin{pmatrix}0\\0\\R_z\end{pmatrix}$
2	B	1	-1	1	-1	1	-1	$\begin{pmatrix}0\\0\\0\end{pmatrix}$	$\begin{pmatrix}0 & 0 & 0\\0 & 0 & 0\\0 & 0 & 0\end{pmatrix}$	$\begin{pmatrix}0\\0\\0\end{pmatrix}$
3	$E_1^{(1)}$	1	ε	$-\varepsilon^*=\varepsilon^2$	$-1=\varepsilon^3$	$-\varepsilon=\varepsilon^4$	$\varepsilon^*=\varepsilon^5$	$\frac{1}{2}\begin{pmatrix}x+iy\\y-ix\\0\end{pmatrix}$	$\frac{1}{2}\begin{pmatrix}0 & 0 & xz+iyz\\0 & 0 & yz-ixz\\zx+izy & zy-izx & 0\end{pmatrix}$	$\frac{1}{2}\begin{pmatrix}R_x+iR_y\\R_y-iR_x\\0\end{pmatrix}$
4	$E_1^{(2)}$	1	$\varepsilon^*=\varepsilon^5$	$-\varepsilon=\varepsilon^4$	$-1=\varepsilon^3$	$-\varepsilon^*=\varepsilon^2$	ε	$\frac{1}{2}\begin{pmatrix}x-iy\\y+ix\\0\end{pmatrix}$	$\frac{1}{2}\begin{pmatrix}0 & 0 & xz-iyz\\0 & 0 & yz+ixz\\zx-izy & zy+izx & 0\end{pmatrix}$	$\frac{1}{2}\begin{pmatrix}R_x-iR_y\\R_y+iR_x\\0\end{pmatrix}$
5	$E_2^{(1)}$	1	$-\varepsilon^*=\varepsilon^2$	$-\varepsilon=\varepsilon^4$	1	$-\varepsilon^*=\varepsilon^2$	$-\varepsilon=\varepsilon^4$	$\begin{pmatrix}0\\0\\0\end{pmatrix}$	$\frac{1}{4}\begin{pmatrix}xx-yy+i(xy+yx) & xy+yx-i(xx-yy) & 0\\xy+yx-i(xx-yy) & -xx+yy-i(xy+yx) & 0\\0 & 0 & 0\end{pmatrix}$	$\begin{pmatrix}0\\0\\0\end{pmatrix}$
6 6	$E_2^{(2)}$	1	$-\varepsilon=\varepsilon^4$	$-\varepsilon^*=\varepsilon^2$	1	$-\varepsilon=\varepsilon^4$	$-\varepsilon^*=\varepsilon^2$	$\begin{pmatrix}0\\0\\0\end{pmatrix}$	$\frac{1}{4}\begin{pmatrix}xx-yy-i(xy+yx) & xy+yx+i(xx-yy) & 0\\xy+yx+i(xx-yy) & -xx+yy+i(xy+yx) & 0\\0 & 0 & 0\end{pmatrix}$	$\begin{pmatrix}0\\0\\0\end{pmatrix}$

Tab.2 Character table of the point group C_6 ($\varepsilon = \exp(2\pi i/6)$, $\varepsilon + \varepsilon^* = 1$, $\varepsilon - \varepsilon^* = i\sqrt{3}$)

The decomposition has been given explicitly for a polar vector (column 4), a second-rank general polar tensor (column 5), and an axial vector (column 6).

2.5 SELECTION RULES

Note a fundamental aspect: an arbitrary tensor transforms as the product of the coordinates indicated by the indices on the tensor components when a linear orthogonal symmetry operation is applied. Selection rules for IR absorption, the linear Raman effect and hyper-Raman effect are deduced from symmetry properties of the crystals in the following way:

The intensity of a Raman line is determined by the matrix elements of the polarizability tensor α, see (1-7) and (1-9). The intensity of an IR absorption band on the other hand is proportional to the square of the matrix element of the electric dipole moment $\underline{P}$. The intensity of a hyperraman line finally depends on the matrix element of the corresponding hyperpolarizability. All these matrix elements have the form

$$\int_{-\infty}^{+\infty} \psi^*_{v'} \, \hat{O} \, \psi_{v''} \, d\tau \quad , \tag{2-10}$$

where $\hat{O}$ stands for the operator of the polarizability, the electric dipole moment, or the hyperpolarizability. For IR absorption and Stokes scattering $\psi_{v'}$ and $\psi_{v''}$ are the eigenfunctions of the upper and lower energy states, respectively. Vibrations of atoms in crystals may in good approximation be described by harmonic oscillators. The corresponding eigenfunctions are determined essentially by the Hermitian polynomials /13/; v' and v" are vibrational quantum numbers. If interaction with light takes place in the ground state (i.e. hot bands are left out of consideration), v" = 0. It can be derived from the Hermitian polynomials that the eigenfunction of the ground state ψ_0 for any normal vibration is totally symmetric for all symmetry operations

$$\hat{R} \, \psi_0 = \psi_0 \quad . \tag{2-11}$$

$\hat{R}$ denotes the operator in question. The representation of ψ_0 is said to be totally symmetric or identical (id)

$$\Gamma(\psi_0) = \Gamma_{id} \quad . \tag{2-12}$$

It can further be derived from the Hermitian polynomials that the eigenfunction of a normal vibration in its first excited state has the same symmetry properties as the normal coordinate Q_j itself

$$\Gamma(\psi^*_{v''}) = \Gamma^{(\gamma)*} \quad . \tag{2-13}$$

The superscript (γ) indicates the symmetry species.

If there are N atoms in the unit cell the 3 N degrees of freedom are distributed within the following types of waves near the center of the 1BZ for the principal directions of the crystal:

1 longitudinal acoustical branch (LA)
2 transverse acoustical branches (TA)
N-1 longitudinal optical branches (LO) and
2N-2 transverse optical branches (TO).

The acoustical branches are neglected in the following considerations. They cause Brillouin scattering. The remaining normal coordinates Q_j therefore are counted as $j = 1,\ldots,3N-3$, and $d\tau$ in (2-10) correspondingly is $d\tau = dQ_1 \cdot dQ_2 \ldots dQ_{3N-3}$.

The matrix element (2-10) becomes different from 0 if the integrand or parts of it are invariant for all symmetry transformations of the point group. In terms of representation theory this means that the matrix element is different from 0 only if the integrand or parts of it are transformed by means of the totally symmetric (identical) representation. A normal vibration thus is 'allowed' if the representation of the integrand (int) contains the identical representation

$$\Gamma_{id} \subset \Gamma_{int} \quad . \tag{2-14}$$

The representation of the integrand is obtained by forming the direct-product representation taking into account (2-12) and (2-13):

$$\Gamma_{int} = \Gamma(\psi^*_{v'}) \times \Gamma(\hat{O}) \times \Gamma(\psi_{v''=0}) = \Gamma^{(\gamma)*} \times \Gamma(\hat{O}) \times \Gamma_{id} \quad . \tag{2-15}$$

In order to decide whether (2-14) holds or not, the character χ_{int} of the integrand is required. Because $\chi_{id}(\hat{R}) = 1$ for all $\hat{R}$,

$$\chi_{int}(\hat{R}) = \chi^{(\gamma)}(\hat{R})^* \cdot \chi_{\hat{O}}(\hat{R}) \quad . \tag{2-16}$$

In general, the representation of the integrand is reducible, i.e. it may be split into irreducible representations. We therefore have to form

$$n_{id} = (1/[P]) \sum_{q=1}^{r} k_q \cdot \chi^{(\gamma)}(\hat{R})^* \cdot \chi_{\hat{O}}(\hat{R}) \quad . \tag{2-17}$$

Eq. (2-14) holds if $n_{id} > 0$ and fails if $n_{id} = 0$. n_{id} indicates how many times the identical representation is contained in the representation of the integrand. $[P]$ denotes the total number of group elements of the point group P (order of the point group in question), k_q stands for the number of group elements in the class q, and r is the total number of classes of group elements in P. $\chi^{(\gamma)}(\hat{R})$ finally denotes the character of the irreducible representation of the symmetry species γ for the symmetry operation $\hat{R}$. $\hat{R}$ is the operator of an arbitrary symmetry element in the class q. All these quantities can be directly obtained from the character tables.

<u>Selection rules for IR absorption</u> can be derived if $\hat{O}$ is replaced by the operator of the electric dipole moment $\underline{P}$. $\underline{P}$ is a polar vector (or first-rank tensor) which is transformed in the same way as the cartesian coordinates $\rho = x, y, z$.

Every symmetry operation $\hat{R}$ can be written as a transformation matrix

$$\hat{R} = \begin{pmatrix} \cos\phi_R & -\sin\phi_R & 0 \\ \sin\phi_R & \cos\phi_R & 0 \\ 0 & 0 & \pm 1 \end{pmatrix} \tag{2-18}$$

if the z axis of the coordinate system is parallel to the most-fold rotational axis or perpendicular to a mirror plane if there is no axis. The different symmetry operations are distinguished by the angles ϕ_R. The positive sign of $R_{zz} = \det\hat{R} = \pm 1$ holds for all symmetry operations derived from a pure rotation (proper symmetry operations), and the negative sign for all those derived from a mirror rotation (improper symmetry operations). The character of the representation which is the trace of the matrix (2-18) can therefore be written as

$$\chi_\mu(R) = 2\cos\phi_R \pm 1 = 2\cos\phi_R + \det\hat{R} \equiv f_R \tag{2-19}$$

where f_R refers to the nomenclature of Behringer /39/. Now the selection rules can be derived explicitly from (2-17). In the point group C_6, for example, only the species A, $E_1^{(1)}$ and $E_1^{(2)}$ are IR-active.

Because the translation $\underline{Q}^T$ is also a polar vector, we obtain the same selection rules. These are of importance for the acoustical modes: Brillouin scattering.

Usually in molecular spectroscopy it is sufficient to decide whether a symmetry species is allowed or not. When studying single crystals, however, we may go further: we can decide which component of $\underline{P}$ is responsible for the IR activity of a certain species. This information is obtained when applying the projection operator /44, 52/

$$\hat{P}_{ii}^{(\gamma)} = (d_\gamma/[P]) \sum_{R \in P} D_{ii}^{(\gamma)}(\hat{R})^* \cdot \hat{R} \quad . \tag{2-20}$$

(The symbol ^ indicates an operator as before.) γ denotes the symmetry species (see e.g. Table 2, column 1), and the indices ii ($i = 1, \ldots, d_\gamma$) the diagonal elements of the representation matrix $D_{ij}^{(\gamma)}(R)$. The dimension d_γ of the representation is identical to the character of the same operation obtained from the E column in the character tables for each symmetry species γ. The sum is taken over all group elements of P.

As an example we again take the point group C_6, which has only one-dimensional representations ($d_\gamma = 1$). The index i can therefore be omitted and $D^{(\gamma)}(\hat{R})^* = \chi^{(\gamma)}(\hat{R})^*$. The projection operator is thus

$$\hat{P}^{(\gamma)} = (1/[P]) \sum_{R \in P} \chi^{(\gamma)}(\hat{R})^* \cdot \hat{R} \quad . \tag{2-21}$$

In order to derive the detailed selection rules for IR absorption and the translations, this operator has to act on each of the coordinates x_ρ ($\rho = 1, 2, 3$). With $x_1 = x$, $x_2 = y$, $x_3 = z$ and $\hat{R}x_\rho = \sum_{\rho'=1}^{3} R_{\rho\rho'} x_{\rho'}$ we get

$$\hat{P}^{(\gamma)} x_\rho = (1/[P]) \sum_{\rho'=1}^{3} x_\rho \sum_{R \in P} \chi^{(\gamma)}(\hat{R})^* \cdot R_{\rho\rho'} \tag{2-22}$$

where $R_{\rho\rho'}$ are the matrix elements of (2-18). The matrices are given explicitly for all symmetry operations of C_6 in Table 3. The totally symmetric species yield

$$\hat{P}^{(1)} x_1 = \hat{P}^{(1)} x_2 = 0 \qquad \text{and} \qquad P^{(1)} x_3 = x_3 \quad .$$

Correspondingly, in column 4 of Table 2 we find the vector

$$\begin{pmatrix} 0 \\ 0 \\ z \end{pmatrix}$$

for the A species. In order to restrict the number of indices, the coordinate system has again been indicated in the table by x, y, z. Our equivalent denominations (x, y, z) and (x_1 x_2 x_3) are identical with the right-angle system (Ox_1, Ox_2, Ox_3) used by Nye /45/.

For the B species:

$$\hat{P}^{(2)} x_1 = \hat{P}^{(2)} x_2 = \hat{P}^{(2)} x_3 = 0 \quad ,$$

for the one-dimensional symmetry species $E_1^{(1)}$ with $\gamma = 3$:

$$\hat{P}^{(3)} x_1 = (1/2)(x_1 + ix_2) \ , \quad \hat{P}^{(3)} x_3 = (1/2)(x_2 - ix_1) \ , \quad \hat{P}^{(3)} x_3 = 0 \quad ,$$

and for the $E_1^{(2)}$ species with $\gamma = 4$:

$$\hat{P}^{(4)} x_1 = (1/2)(x_1 - ix_2) \ , \quad \hat{P}^{(4)} x_2 = (1/2)(x_2 + ix_1) \ , \quad \hat{P}^{(4)} x_3 = 0 \quad .$$

The vectors in column 4 of Table 2 summarize these results. Finally, for the symmetry species $E_2^{(1)}$ and $E_2^{(2)}$ the projection operator acting on all coordinates gives 0.

Symmetry species with the complex conjugate characters of irreducible representations are degenerate. Thus in many tables (e.g. those given in /13, 14, 29/) the species $E_1^{(1)}$, $E_1^{(2)}$ and $E_2^{(1)}$, $E_2^{(2)}$ are not distinguished from each other. Usually, the selection rules are given only in the well-known comprehensive form, e.g. (x_1, x_2) or (T_x, T_y) for E_1. Rigorously, this is not correct because the symmetry species E_1 and E_2 are reducible in the complex space.

IR-active species are also referred to as 'polar' and the corresponding vibrational modes consequently as 'polar modes'. We shall

	E	C_6	$C_6^2 = C_3$	$C_6^3 = C_2$	$C_6^4 = C_3^2$	$C_6 = \bar{C}_6$
$\hat{R}$	$\begin{pmatrix} 1 & 0 & 0 \\ 0 & 1 & 0 \\ 0 & 0 & 1 \end{pmatrix}$	$\begin{pmatrix} \frac{1}{2} & -\frac{\sqrt{3}}{2} & 0 \\ +\frac{\sqrt{3}}{2} & \frac{1}{2} & 0 \\ 0 & 0 & 1 \end{pmatrix}$	$\begin{pmatrix} -\frac{1}{2} & -\frac{\sqrt{3}}{2} & 0 \\ +\frac{\sqrt{3}}{2} & -\frac{1}{2} & 0 \\ 0 & 0 & 1 \end{pmatrix}$	$\begin{pmatrix} -1 & 0 & 0 \\ 0 & -1 & 0 \\ 0 & 0 & 1 \end{pmatrix}$	$\begin{pmatrix} -\frac{1}{2} & +\frac{\sqrt{3}}{2} & 0 \\ -\frac{\sqrt{3}}{2} & -\frac{1}{2} & 0 \\ 0 & 0 & 1 \end{pmatrix}$	$\begin{pmatrix} \frac{1}{2} & +\frac{\sqrt{3}}{2} & 0 \\ -\frac{\sqrt{3}}{2} & \frac{1}{2} & 0 \\ 0 & 0 & 1 \end{pmatrix}$

Tab.3 Transformation matrices for the symmetry operations $\hat{R}$ of the point group C_6 ($\cos 2\pi/6 = 0.5$ and $\sin 2\pi/6 = \sqrt{3}/2$).

see later (3.1 to 3.3) that polaritons are associated only with polar modes.

In order to derive the selection rules for the linear Raman effect, $\hat{O}$ has to be replaced by the polarizability operator in (2-10), (2-16) and (2-17). The transformation of a polar second-rank symmetric tensor /45/ requires the character of the representation to be /13/

$$\chi_\alpha(R) = 2\cos\phi_R(\pm 1 \pm 2\cos\phi_R)$$

$$= 2\det\hat{R}\cos\phi_R(1 + 2\cos\phi_R) \equiv f''_R \quad . \qquad (2\text{-}23)$$

The same sign rule holds as for (2-19). Taking (2-17) into account, the general condition (2-14) can be discussed again. In order to derive detailed selection rules, the projection operator (2-20) or (2-21) now has to act on the products $x_\rho \cdot x_\sigma$ (ρ,σ = 1, 2, 3). For one-dimensional species

$$\hat{P}^{(\gamma)} x_\rho x_\sigma = (1/[P]) \sum_{\rho',\sigma'=1}^{3} x_{\rho'} x_{\sigma'} \sum_{R \in P} \chi^{(\gamma)}(\hat{R})^* \cdot R_{\rho\rho'} R_{\sigma\sigma'} \quad . \qquad (2\text{-}24)$$

The sum over ρ' and σ' contains nine terms with the index combinations 11, 22, 33, 12, 21, 13, 31, 23 and 32. No presupposition is claimed for tensors derived in this way.

The application to point group C_6 now becomes

$$\hat{P}^{(3)} x_1 x_3 = (1/2)(x_1x_3 + ix_2x_3) \quad , \quad \hat{P}^{(3)} x_2 x_3 = (1/2)(x_2x_3 - ix_1x_3) \quad , \text{ etc.}$$

When acting on the products x_1x_1, x_2x_2, x_3x_3, x_1x_2 and x_2x_1, the operator $\hat{P}^{(3)}$ gives 0. The result obtained for the symmetry species $E_1^{(1)}$ has been given in tensor form in column 5 of Table 2. Detailed selection rules for the other symmetry species are derived in the same way.

We point out some characteristic properties of the tensors:

1) If all tensors are added up, the tensor $(x_{\rho'}, x_{\sigma'})$ with ρ',σ' = 1, 2, 3 is again obtained. This may advantageously be used to check the calculations. We have only considered $(x_{\rho'}, x_{\sigma'})$ because it shows the same symmetry properties as the

polarizability tensor ($\alpha_{\rho\sigma}$). For instance, the tensor for A-type vibrations from column 5 in Table 2 can be written more explicitly:

$$(1/2) \begin{pmatrix} \alpha_{xx}+\alpha_{yy} & \alpha_{xy}-\alpha_{yx} & 0 \\ -(\alpha_{xy}-\alpha_{yx}) & \alpha_{xx}+\alpha_{yy} & 0 \\ 0 & 0 & 2\alpha_{zz} \end{pmatrix} \quad . \qquad (2\text{-}25)$$

Such tensors represent a decomposition of the polarizability tensor and are transformed in the same way as the matrices of the corresponding representations. For C^6 with only one-dimensional symmetry species, the representations are reduced to the characters of the irreducible representations.

2) As can easily be seen, the tensors of the unpolar symmetry species $E_2^{(1)}$ and $E_2^{(2)}$ are symmetric

$$\alpha_{\rho\sigma} = \alpha_{\sigma\rho} \quad . \qquad (2\text{-}26)$$

3) Furthermore the tensor components of the two species $E_1^{(1)}$ and $E_1^{(2)}$, or $E_2^{(1)}$ and $E_2^{(2)}$, respectively, are complex conjugate

$$(x_{\rho'} x_{\sigma'})_{E_j^{(2)}} = (x_{\rho'} x_{\sigma'})^*_{E_j^{(1)}} \quad \text{or} \quad (\alpha_{\rho\sigma})_{E_j^{(2)}} = (\alpha_{\rho\sigma})^*_{E_j^{(1)}} \ , \qquad (2\text{-}27)$$

$j = 1, 2.$

This can be seen directly from (2-24). (The characters are complex conjugate.)

4) The tensor of the A species finally contains real elements only and an antisymmetric part because

$$x_{\sigma'} x_{\rho'} = -x_{\rho'} x_{\sigma'} \quad \text{or} \quad \alpha_{\sigma\rho} = -\alpha_{\rho\sigma} \quad \text{for} \quad \rho \neq \sigma \quad . \qquad (2\text{-}28)$$

All common character tables giving information on the Raman activity state that the polarizability tensor is symmetric. Such tables, for instance, contain no information on the off-diagonal tensor elements for A species. In practice the polarizability tensors may contain nonzero antisymmetric parts for resonance Raman scattering /46, 53, 54/ or when the electronic ground state is degenerate /55, 56/. We denote the antisymmetric part by $\{\alpha\}$.

It can be shown /47/ that for $\{\alpha\}$

$$\chi_{\{\alpha\}}(\hat{R}) = 1 + 2\det\hat{R} \cdot \cos\phi_R = \det\hat{R} \cdot f_R = f'_R \qquad (2\text{-}29)$$

holds instead of (2-23). Eq.(2-29) is identical with the character of a molecular rotation /48/ and can, consequently, be written

$$\chi_{\{\alpha\}}(\hat{R}) = \chi_{Q^R}(\hat{R}) \quad . \qquad (2\text{-}30)$$

(Note that the suffix R on Q means 'rotation' whereas $\hat{R}$ in brackets stands for all symmetry operations.)

The <u>antisymmetric part of the polarizability tensor</u> transforms as the normal coordinate vector $\underline{Q}^R$ for rotating nonlinear molecules. $\underline{Q}^R$ defining the rotational momentum is an axial vector. For proper symmetry operations it transforms as a polar vector and for improper operations as a polar vector in the opposite direction /48/. In order to derive the selection rules for rotations, a factor (-1) has thus to be added in terms referring to the improper symmetry operations when the projection operator (2-20) or (2-21) is applied. The selection rules for $\{\alpha\}$ may be derived directly, either by examining the distribution of the polarizability tensor on its antisymmetric parts by means of the projection operator applied to the products $x_\rho\, x_\sigma$,or by calculating the transformation properties of the rotation from (2-29). In the latter case /47/

$$\begin{pmatrix} Q^R_{x_1} \\ Q^R_{x_2} \\ Q^R_{x_3} \end{pmatrix} = \begin{pmatrix} R_x \\ R_y \\ R_z \end{pmatrix} = \begin{pmatrix} \{\alpha_{x_2x_3}\} \\ \{\alpha_{x_3x_1}\} \\ \{\alpha_{x_1x_2}\} \end{pmatrix} = (1/2) \begin{pmatrix} \alpha_{yz}-\alpha_{zy} \\ \alpha_{zx}-\alpha_{xz} \\ \alpha_{xy}-\alpha_{yx} \end{pmatrix} \quad . \qquad (2\text{-}31)$$

This implies that e.g. the antisymmetric component $\{\alpha_{x_2x_3}\}$ transforms in the same way as the x_1 component of the rotational normal coordinate $\underline{Q}^R$. For the point group C_6, consequently, a rotation around the x_3 axis $Q^R_{x_3} = R_z$ is allowed and the antisymmetric component $(\alpha_{xy}-\alpha_{yx})/2$ of the A species does not vanish. Furthermore, because C_6 contains only proper symmetry operations, the selection rules for rotations are identical to those of translations and the electric dipole moment.

5) The tensors for the $E_1^{(1)}$ and $E_1^{(2)}$ species in column 5 of Table 2 are nonhermitian. They contain an antisymmetric part which for a second-rank tensor is given by

$$\{\alpha_{\rho\sigma}\} = (\alpha_{\rho\sigma}-\alpha_{\sigma\rho})/2 \quad . \tag{2-32}$$

The projection operator (2-20) can be used as before to derive this and the corresponding relations for tensors of higher order. The sum covers the elements of the permutation group of second or higher order (see /59/).

According to (2-32) the nonsymmetrical part of $\{x_2x_3\}$ in, e.g. the symmetry species $E_1^{(1)}$, becomes

$$\{x_2x_3\} = \left[yz-zy-i(xz-zx)\right]/4 \quad , \tag{2-33}$$

which because of (2-31) and (2-32) corresponds to the rotational normal coordinate component $(R_x+iR_y)/2$. The antisymmetric parts of the polarizability tensor and the corresponding components of the rotational coordinate for the point group C_6 are listed in full in Table 4.

The decomposition of an arbitrary rank tensor ϕ for more-dimensional symmetry species has to be performed in the following way.

A tensor $\phi_1^{(\gamma)}$ transforming as the irreducible representation γ has first to be determined by the projection operator (2-20). E.g., for i = 1

$$\hat{P}_{11}^{(\gamma)}\phi = (d_\gamma/[P])\sum_{R\epsilon P} D_{11}^{(\gamma)}(R)^{*}\cdot\hat{R}\phi = \phi_1^{(\gamma)} \tag{2-34}$$

holds. All components of ϕ have to be calculated in this way. In the symmetry species γ (of dimension d_γ), however, there are $(d_\gamma-1)$ additional tensors which may be derived by means of the transfer (or shift) operator

C_6	$\{\alpha_{\rho\sigma}\}$	Q_ρ^R
A	$\{x_2x_3\} = \{x_3x_1\} = 0$ $\{x_1x_2\} = \frac{1}{2}(xy-yx) = R_z$	$\begin{pmatrix} 0 \\ 0 \\ R_z \end{pmatrix}$
$E_1^{(1)}$	$\{x_2x_3\} = \frac{1}{4}[yz-zy-i(xz-zx)] =$ $= \frac{1}{2}(R_x+iR_y)$ $\{x_3x_1\} = \frac{1}{4}[zx-xz+i(zy-yz)] =$ $= \frac{1}{2}(R_y-iR_x)$ $\{x_1x_2\} = 0$	$\frac{1}{2}\begin{pmatrix} R_x+iR_y \\ R_y-iR_x \\ 0 \end{pmatrix}$
$E_1^{(2)}$	$\{x_2x_3\} = \frac{1}{4}[yz-zy+i(xz-zx)] =$ $= \frac{1}{2}(R_x-iR_y)$ $\{x_3x_1\} = \frac{1}{4}[zx-xz-i(zy-yz)] =$ $= \frac{1}{2}(R_y+iR_x)$ $\{x_1x_2\} = 0$	$\frac{1}{2}\begin{pmatrix} R_x-iR_y \\ R_y+iR_x \\ 0 \end{pmatrix}$

Tab.4 The antisymmetric part of the polarizability tensor $\{\alpha_{\rho\sigma}\}$, see Table 2, and the corresponding rotational normal coordinate vector for the point group C_6. $\{\rho\sigma\}$ is used as an abbreviation for $\{\alpha_{\rho\sigma}\}$; $\rho,\sigma = x_1,x_2,x_3$.

$$\hat{P}_{ji}^{(\gamma)} = (d_\gamma/[\mathcal{P}]) \sum_{R\in\mathcal{P}} D_{ji}^{(\gamma)}(R)^*\hat{R} \quad , \qquad j = 1,\ldots,(d_\gamma-1) \quad , \tag{2-35}$$

see, for instance, /52/ or /44/. Applied to $\phi_1^{(\gamma)}$, the transfer operator determines

$$\hat{P}_{j1}^{(\gamma)}\phi_1^{(\gamma)} = \phi_j^{(\gamma)} \quad . \tag{2-36}$$

We illustrate the method on the smallest non-abelian point group $\mathcal{D}_3$. The matrices of all its symmetry operations are given in Table 5. A decomposition of ϕ can be achieved if all symmetry operations are known. The two generating elements are C_3 and C_2^x /44/. The remaining group elements can be constructed in the usual manner. For more-dimensional species it is not sufficient just to know the character of the irreducible representations; the total representation matrices are required. These have been tabulated explicitly for the generating elements by Poulet and Mathieu /33/. According to (2-34) and Table 5, the components of a vector (IR absorption, translation, rotation) become

$$\hat{P}_{11}^{(3)}x_1 = x_1 \qquad \text{and} \qquad \hat{P}_{11}^{(3)}x_2 = \hat{P}_{11}^{(3)}x_3 = 0 \quad ,$$

and

$$\phi_1^{(3)} = \begin{pmatrix} x_1 \\ 0 \\ 0 \end{pmatrix} \quad . \tag{2-37}$$

From (2-36) there follows the general relation

$$\hat{P}_{21}^{(3)}x_\rho = (2/6)x_1 \sum_{R\in\mathcal{P}} D_{21}^{(3)}(R)\cdot R_{\rho 1}$$

which in our example determines

$$\hat{P}_{21}^{(3)}x_1 = \hat{P}_{21}^{(3)}x_3 = 0 \quad \text{and} \quad \hat{P}_{21}^{(3)}x_2 = x_1 \quad ,$$

i.e.

$$\phi_2^{(3)} = \begin{pmatrix} 0 \\ x_1 \\ 0 \end{pmatrix} \quad . \tag{2-38}$$

1	2	3						4	5	6
γ	$\mathcal{D}_3$	$\begin{pmatrix}1&0&0\\0&1&0\\0&0&1\end{pmatrix}$	$\begin{pmatrix}-\frac{1}{2}&-\frac{\sqrt{3}}{2}&0\\\frac{\sqrt{3}}{2}&-\frac{1}{2}&0\\0&0&1\end{pmatrix}$	$\begin{pmatrix}-\frac{1}{2}&\frac{\sqrt{3}}{2}&0\\-\frac{\sqrt{3}}{2}&-\frac{1}{2}&0\\0&0&1\end{pmatrix}$	$\begin{pmatrix}1&0&0\\0&-1&0\\0&0&-1\end{pmatrix}$	$\begin{pmatrix}-\frac{1}{2}&-\frac{\sqrt{3}}{2}&0\\-\frac{\sqrt{3}}{2}&\frac{1}{2}&0\\0&0&-1\end{pmatrix}$	$\begin{pmatrix}-\frac{1}{2}&\frac{\sqrt{3}}{2}&0\\\frac{\sqrt{3}}{2}&\frac{1}{2}&0\\0&0&-1\end{pmatrix}$	μ_ρ Q^T_ρ	$\alpha_{\rho\sigma}$	Q^R_ρ
		E	C_3^z	$\overline{C_3^z}$	C_2^x	$C_2^{x-60^\circ}=C_2'$	$C_2^{x+60^\circ}=C_2''$			$\{\alpha_{\rho\sigma}\}$
	=32	1	3	3	2	2	2			see (2-32)
1	A_1	1	1	1	1	1	1	$\begin{pmatrix}0\\0\\0\end{pmatrix}$	$\frac{1}{2}\begin{pmatrix}xx+yy&&\\&xx+yy&\\&&2zz\end{pmatrix}$	$\begin{pmatrix}0\\0\\0\end{pmatrix}$
2	A_2	1	1	1	-1	-1	-1	$\begin{pmatrix}0\\0\\z\end{pmatrix}$	$\frac{1}{2}\begin{pmatrix}&xy-yx&\\-xy+yx&&\\&&\end{pmatrix}$	$\begin{pmatrix}0\\0\\R_z\end{pmatrix}$
3	$E^{(1)}$	$\begin{pmatrix}1&0\\0&1\end{pmatrix}$	$\begin{pmatrix}-\frac{1}{2}&-\frac{\sqrt{3}}{2}\\\frac{\sqrt{3}}{2}&-\frac{1}{2}\end{pmatrix}$	$\begin{pmatrix}-\frac{1}{2}&\frac{\sqrt{3}}{2}\\-\frac{\sqrt{3}}{2}&-\frac{1}{2}\end{pmatrix}$	$\begin{pmatrix}1&0\\0&-1\end{pmatrix}$	$\begin{pmatrix}-\frac{1}{2}&-\frac{\sqrt{3}}{2}\\-\frac{\sqrt{3}}{2}&\frac{1}{2}\end{pmatrix}$	$\begin{pmatrix}-\frac{1}{2}&\frac{\sqrt{3}}{2}\\\frac{\sqrt{3}}{2}&\frac{1}{2}\end{pmatrix}$	$\begin{pmatrix}x\\0\\0\end{pmatrix}$	$\frac{1}{2}\begin{pmatrix}xx-yy&&\\&-xx+yy&2yz\\&2zy&\end{pmatrix}$	$\begin{pmatrix}R_x\\0\\0\end{pmatrix}$
	$E^{(2)}$							$\begin{pmatrix}0\\y=x\\0\end{pmatrix}$	$\frac{1}{2}\begin{pmatrix}&-xx+yy&-2yz\\-xx+yy&&\\-2zy&&\end{pmatrix}$	$\begin{pmatrix}0\\R_y=R_x\\0\end{pmatrix}$

Tab.5 Matrices of the irreducible representations for all symmetry operations of the point groups $\mathcal{D}_3$.

The xy plane consequently is isotropic and crystals belonging to direction group $\mathcal{D}_3$ are uniaxial.

Furthermore, from (2-34) it follows for the components of a second-rank tensor corresponding to the linear Raman effect or two-photon absorption that

$$\hat{P}_{11}^{(3)} x_\rho x_\sigma = (1/3) \sum_{\rho',\sigma'=1}^{3} x_{\rho'} x_{\sigma'} \sum_{R\in\mathcal{D}_3} D_{11}^{(3)}(R) \cdot R_{\rho\rho'} R_{\sigma\sigma'} \quad .$$

In our example we explicitly get

$$\hat{P}_{11}^{(3)} x_1x_1 = -\hat{P}_{11}^{(3)} x_2x_2 = (xx-yy)/2 \; ; \quad \hat{P}_{11}^{(3)} x_2x_3 = yz \; ; \quad \hat{P}_{11}^{(3)} x_3x_2 = zy$$

$$\hat{P}_{11}^{(3)} x_3x_3 = \hat{P}_{11}^{(3)} x_1x_2 = \hat{P}_{11}^{(3)} x_2x_1 = \hat{P}_{11}^{(3)} x_1x_3 = \hat{P}_{11}^{(3)} x_3x_1 = 0 \quad .$$

These components form the tensor $(\alpha_{\rho\sigma})$ in the line $E^{(1)}$ of Table 5. The polarizability tensor for the $E^{(2)}$ species, on the other hand, can be calculated using the transfer operator (2-35)

$$\hat{P}_{21} x_\rho x_\sigma = (1/3) \sum_{\rho',\sigma'=1}^{3} x_{\rho'} x_{\sigma'} \sum_{R\in P} D_{21}^{(3)}(R) \cdot R_{\rho\rho'} R_{\sigma\sigma'} \quad .$$

$x_{\rho'}$ and $x_{\sigma'}$ have to be replaced by the corresponding components of the polarizability tensor $(\alpha_{\rho\sigma})$. Explicitly, we get

$$\hat{P}_{21}^{(3)} x_1x_2 = (1/3)\left[(1/2)(xx-yy)(-1/2)+(1/2)(-xx+yy)(1/2)\right] =$$

$$= (1/2)(-xx+yy) \quad ,$$

$$\hat{P}_{21}^{(3)} x_2x_1 = (1/2)(-xx+yy) \; ; \quad \hat{P}_{21}^{(3)} x_1x_3 = -yz \; ; \quad \hat{P}_{21}^{(3)} x_3x_1 = -zy \quad ,$$

$$\hat{P}_{21}^{(3)} x_\ell x_\ell = 0 \quad \text{for} \quad \ell = 1,\,2,\,3 \; ; \quad \hat{P}_{21}^{(3)} x_2x_3 = \hat{P}_{21}^{(3)} x_3x_2 = 0 \quad .$$

According to (2-31) and (2-32), the antisymmetric parts of the tensors derived in this way lead to the selection rules for rotations given in column 6 of Table 5.

The decomposition of first-, second-, third-, and fourth-rank tensors for all point groups playing a role as crystal classes have been calculated by Winter and Brandmüller /57/. The results obtained for second-rank tensors are of importance also for nonlinear two-photon absorptions, see above /56/. The decomposition of third-rank tensors leads to the selection rules of the β-hyperraman effect, which is a three-photon process, see (1-13) and /35/. Finally, the selection rules for stimulated Raman scattering and the γ-hyperraman effect (four-photon processes) are determined by a fourth-rank tensor. Corresponding references are given in /35/. From such tables as Tables 2 and 4 the selection rules for translations, rotations, IR absorption, and the linear Raman effect can be read off. The normal coordinates of type B are neither IR active nor observable by spontaneous Raman scattering. They are referred to as 'silent modes'. In hyperraman scattering, however, they are not forbidden. General selection rules for the hyperraman effect are derived when $\hat{O}$ in (2-17) is replaced by the operator of the hyperpolarizability $\hat{\beta}$:

$$\chi_{\beta}(\hat{R}) = 2\cos\phi_R(4\cos^2\phi_R + 2\det\hat{R}\cos\phi_R - 1) \quad . \tag{2-39}$$

Detailed selection rules again are obtained when applying the projection operator. Results have been published for the tensor $(\beta_{\rho\sigma\tau})$, for its antisymmetric part, and for a fourth-rank tensor $(\gamma_{\rho\sigma\tau\upsilon})$ by Christie and Lockwood /49/ and Menzies /50/. The degenerate species, which can be reduced to a one-dimensional representation have not, however, been explicitly separated.

The selection rules so far derived by group-theoretical methods determine the fundamental normal vibrations allowed in IR absorption or in the linear Raman effect. Selection rules for the harmonics and combinations can be calculated by similar methods, see /13/.

In molecular spectroscopy it is usually sufficient to know whether a certain normal coordinate is allowed or not. Assignments of totally symmetric species can be achieved when analyzing the depolarization ratio. Assignments of other symmetry species, however, can be carried out only on single crystals with the use of well-defined scattering geometries. We therefore discuss the Raman tensors in more detail.

Additional Literature

Birman, J.L.: Theory of Multiple-Dipole Resonance Raman Scattering by Phonons /319/.

2.6 THE RAMAN TENSOR

The form of the polarizability tensor for the A species is derived as illustrated for the point group C_6 in (2-25). Except for some special cases cited above, the intensity of a Raman line is determined only by the symmetric part of this tensor. The different matrix elements (2-10) of the polarizability components are required numerically for (1-7). The following abbreviations are introduced for the matrix elements:

$$\int_{-\infty}^{+\infty} \psi^*_{v'}(1/2)(\alpha_{xx}+\alpha_{yy})\psi_{v''}\, d\tau = a$$

and

$$\int_{-\infty}^{+\infty} \psi^*_{v'}\alpha_{zz}\psi_{v''}\, d\tau = b \quad . \tag{2-40}$$

The symmetric part of the corresponding tensor S for A(z) species in point group C_6 then can be written

$$S^{A(z)} = \begin{pmatrix} a & & \\ & a & \\ & & b \end{pmatrix} \quad . \tag{2-41}$$

The cartesian coordinate added to the symbol of the symmetry species in brackets refers to the IR activity indicated in column 4 of Table 2. The vector $\begin{pmatrix} 0 \\ 0 \\ z \end{pmatrix}$ is simply replaced by (z). The polarizability tensors of the remaining species of C_6 are complex. We shall discuss them later.

In point group D_3 all polarizability components are real, see Table 5. The corresponding Raman tensors S can therefore easily be obtained. The form is identical to (2-41). A_2 modes are represented only by an antisymmetric tensor. These modes therefore are forbidden in first order linear Raman scattering. The tensors of the $E^{(1)}$ and $E^{(2)}$ species contain an antisymmetric part, which must be separated from the symmetric ones. Coordinate combinations such as yz in abbreviation stand for α_{yz}. In general, $\alpha_{yz} \neq \alpha_{zy}$. The decomposition into a symmetric and an antisymmetric part becomes

$$\frac{1}{2}\begin{pmatrix} xx-yy & 0 & 0 \\ 0 & -xx+yy & 2yz \\ 0 & 2yz & 0 \end{pmatrix} = \frac{1}{2}\begin{pmatrix} xx-yy & 0 & 0 \\ 0 & -xx+yy & yz+zy \\ 0 & yz+zy & 0 \end{pmatrix} + \frac{1}{2}\begin{pmatrix} 0 & 0 & 0 \\ 0 & 0 & yz-zy \\ 0 & zy-yz & 0 \end{pmatrix}$$

for $E^{(1)}$ species, and

$$\frac{1}{2}\begin{pmatrix} 0 & -xx+yy & -2yz \\ -xx+yy & 0 & 0 \\ -2zy & 0 & 0 \end{pmatrix} = \frac{1}{2}\begin{pmatrix} 0 & -xx+yy & -(yz+zy) \\ -xx+yy & 0 & 0 \\ -(zy+yz) & 0 & 0 \end{pmatrix} +$$

$$+ \frac{1}{2}\begin{pmatrix} 0 & 0 & -(yz-zy) \\ 0 & 0 & 0 \\ -(zy-yz) & 0 & 0 \end{pmatrix}$$

for $E^{(2)}$ species. Using abbreviations in analogy to (2-40), the symmetric part for the $E^{(1)}(x)$ species is

$$S^{E^{(1)}}(x) = \begin{pmatrix} c & & \\ & -c & d \\ & d & \end{pmatrix} ,$$

and correspondingly

$$S^{E^{(2)}}(y) = \begin{pmatrix} & -c & -d \\ -c & & \\ -d & & \end{pmatrix} . \qquad (2\text{-}42)$$

These tensors are usually referred to as 'Raman tensors'. They have been tabled for all crystal classes by various authors. Poulet and Mathieu /33/ indicate them by $P_{\alpha\beta}((i),n)$. Loudon correspondingly uses the symbol $R^{\tau}_{\sigma\rho}$ in his table /31/. Ovander /58/ has published tables which, however, are not complete. The table published by Mc Clain /56/ is primarily derived for two-photon absorption processes. It also holds, however, for two-photon Raman scattering. In this table, furthermore, the antisymmetric parts are given explicitly.

In symmetry species with complex characters of the irreducible representations the polarizability tensors consequently become complex too. Real tensors may be calculated on the basis of the complex ones in the following way. According to Table 2, the symmetric parts of the polarizability tensors of, e.g. the $E_1^{(1)}$ and $E_1^{(2)}$ species are

$$\frac{1}{4}\begin{pmatrix} & & (xz+zx)+i(yz+zy) \\ & & (zy+yz)-i(zx+xz) \\ (zx+xz)+i(zy+yz) & (zy+yz)-i(zx+xz) & \end{pmatrix} , \qquad (2\text{-}43)$$

and its complex conjugate, respectively. When introducing abbreviations for the matrix elements

$$(1/4)\int_{-\infty}^{+\infty} \psi_{v'}(\alpha_{zx}+\alpha_{xz})\psi_{v''}\, d\tau = c/\sqrt{2} ,$$

and

$$(1/4)\int_{-\infty}^{+\infty} \psi_{v'}(\alpha_{zy}+\alpha_{yz})\psi_{v''}\, d\tau = d/\sqrt{2} \qquad (2\text{-}44)$$

the complex scattering tensors are given by

$$S^{E_1^{(1)}} = (1/\sqrt{2})\begin{pmatrix} & & c+id \\ & & d-ic \\ c+id & d-ic & \end{pmatrix}$$

and

$$S^{E_1^{(2)}} = (1/\sqrt{2})\begin{pmatrix} & & c-id \\ & & d+ic \\ c-id & d+ic & \end{pmatrix} . \qquad (2\text{-}45)$$

Also the decomposition of a polar vector with respect to the irreducible representations requires complex vectors for these two-symmetry species. Thus, according to column 4 of Table 2,

$$E_1^{(1)} : (1/2)\begin{pmatrix} x+iy \\ y-ix \\ 0 \end{pmatrix} = (1/\sqrt{2})(x+iy)(1/\sqrt{2})\begin{pmatrix} 1 \\ -i \\ 0 \end{pmatrix} = r_1\underline{e}_1 \quad ,$$

$$E_1^{(2)} : (1/2)\begin{pmatrix} x-iy \\ y+ix \\ 0 \end{pmatrix} = (1/\sqrt{2})(x-iy)(1/\sqrt{2})\begin{pmatrix} 1 \\ i \\ 0 \end{pmatrix} = r_2\underline{e}_2 \quad ,$$

and

$$A : \begin{pmatrix} 0 \\ 0 \\ z \end{pmatrix} = z\begin{pmatrix} 0 \\ 0 \\ 1 \end{pmatrix} = r_3\underline{e}_3 \quad . \tag{2-46}$$

The magnitudes of the basic vectors have been normalized

$$\underline{e} = (\underline{e}_1\underline{e}_2\underline{e}_3) = (1/\sqrt{2}) \left[\begin{pmatrix} 1 \\ -i \\ 0 \end{pmatrix}\begin{pmatrix} 1 \\ i \\ 0 \end{pmatrix}\begin{pmatrix} 0 \\ 0 \\ \sqrt{2} \end{pmatrix}\right] \quad . \tag{2-47}$$

The corresponding coordinates (which are identical to circular coordinates) are still complex

$$\underline{r} = \begin{pmatrix} (1/\sqrt{2}) & (x+iy) \\ (1/\sqrt{2}) & (x-iy) \\ & z \end{pmatrix} \quad . \tag{2-48}$$

The Raman tensors, however, are expected to provide assignments for experimentally recorded phonon spectra. Consequently, real basis vectors and real coordinates are required. A similarity transformation matrix T may determine the new real basis /44/

$$\underline{e}' = \underline{e}\ T \tag{2-49}$$

which for experimental reasons forms the unit vectors of a rectangular (cartesian) coordinate system. Eq. (2-49) explicitly becomes

$$\left[\begin{pmatrix} 1 \\ 0 \\ 0 \end{pmatrix}\begin{pmatrix} 0 \\ 1 \\ 0 \end{pmatrix}\begin{pmatrix} 0 \\ 0 \\ 1 \end{pmatrix}\right] = (1/\sqrt{2}) \left[\begin{pmatrix} 1 \\ -i \\ 0 \end{pmatrix}\begin{pmatrix} 1 \\ i \\ 0 \end{pmatrix}\begin{pmatrix} 0 \\ 0 \\ \sqrt{2} \end{pmatrix}\right] \begin{pmatrix} T_{11} & T_{12} & T_{13} \\ T_{21} & T_{22} & T_{23} \\ T_{31} & T_{32} & T_{33} \end{pmatrix} \quad . \tag{2-50}$$

Accordingly the unitary transformation matrix is

$$T = (1/\sqrt{2}) \begin{pmatrix} 1 & i & 0 \\ 1 & -i & 0 \\ 0 & 0 & \sqrt{2} \end{pmatrix} \quad . \tag{2-51}$$

The coordinates are transformed by the relation /44/

$$\underline{r}' = T^{-1}\underline{r} = T^{+}\underline{r} \quad , \tag{2-52}$$

where T^{+} denotes the conjugated transposed matrix. From (2-48) and (2-51) the cartesian coordinates

$$\underline{r}' = \begin{pmatrix} x \\ y \\ z \end{pmatrix} \tag{2-53}$$

are obtained. The Raman tensor S for polar modes is a third-rank tensor with components $S_{\tau,\rho\sigma}$, where ρ, σ, τ = x, y, z. We can write this tensor as a three-dimensional run vector with components forming three second-rank tensors

$$\underline{S} = (S_{x,\rho\sigma}\,, S_{y,\rho\sigma}\,, S_{z,\rho\sigma}) \quad . \tag{2-54}$$

According to (2-49) this vector again transforms as

$$\underline{S}' = \underline{S}\;T \quad . \tag{2-55}$$

Using (2-41), (2-45) and (2-51) we explicitly obtain

$$\underline{S}' = \left[\begin{pmatrix} & & c \\ & & d \\ c & d & \end{pmatrix}, \begin{pmatrix} & & -d \\ & & c \\ -d & c & \end{pmatrix}, \begin{pmatrix} a & & \\ & a & \\ & & b \end{pmatrix} \right] \quad . \tag{2-56}$$

This is the real form of the Raman tensor as usually given in the literature /31, 33/, see Appendix 4. There are some discrepancies because different coordinate axes have been chosen by the different authors. Loudon strictly refers to Nye /45/. The table of Poulet and Mathieu differs from Loudon's in the monoclinic crystal class and for D_{3h} and C_{3v}. Finally, there are some minor errors in most of the tables published hitherto. We therefore include a corrected list (Appdx 4) recalculated in the way described above. Winter and Brandmüller /57/ have calculated the tensors for the hyper-Raman effect, which are of importance for interpretations of single-crystal spectra. Corresponding experiments have been performed successfully in recent years, see for instance /49/.

Additional Literature

Birman, J.L., Berenson, Rh.: Scattering tensors and Clebsch-Gordan coefficients in crystals /320/.

Birman, J.L.: Scattering tensors for 'forbidden' resonance Raman scattering in cubic crystals /321/.

Birman, J.L.: Theory of Crystal Space Groups and Infra-Red and Raman Lattice Processes of Insulating Crystals /322/.

3. Dispersion of Polar Optical Modes in Cubic Diatomic Crystals

3.1 HUANG'S EQUATIONS

Phonon dispersion curves $\omega = \omega(\underline{k})$ of diatomic cubic crystals, for instance alkali halides, may be calculated from rigid-ion models, shell models, or breathing-shell models. These models are suitable to describe lattice waves from the edge of the first Brillouin zone (1BZ) at $k \approx 10^8\ \mathrm{cm}^{-1}$ down to $k \gtrsim 10^7\ \mathrm{cm}^{-1}$, a region where the theory can be experimentally verified by neutron scattering. For $k << 10^7\ \mathrm{cm}^{-1}$ in the central part of the 1BZ, no dispersion of optical phonons is predicted by these models, i.e. $\omega(\underline{k}) = \mathrm{const}$. The phase-velocity of all optical phonons vanishes because $\partial\omega/\partial k = 0$. In the simplest such model (the linear diatomic chain) proposed by Born and von Kármán in 1912 /60/, the oscillating atoms are regarded as two different types of mass points. We shall discuss this model in the light of the polariton theory and thus assume the masses to be charged simultaneously. If the masses of anions are indicated by $m_\ominus$ and those of cations by $m_\oplus$, the equations of motion are

$$m_\ominus \ddot{u}_{2n} = f\left[(u_{2n+1} - u_{2n} + a) - (u_{2n} - u_{2n-1} + a)\right] \tag{3-1}$$

and

$$m_\oplus \ddot{u}_{2n-1} = f\left[(u_{2n} - u_{2n-1} + a) - (u_{2n-1} - u_{2n-2} + a)\right] \quad , \tag{3-2}$$

where a denotes the distance between cations and anions and 2a the magnitude of the unit cell. The model implies only nearest-neighbor interaction in the harmonic approximation. f is the force constant. The lattice-displacement coordinates u_i of cations have odd numbers and those of anions even numbers. An ansatz of plane-wave solutions for the u_{2n-1} and u_{2n}, respectively, leads to the prediction of the well-known acoustic and optical dispersion branches. The former exhibit zero energies for k = 0 which, for real crystals, corresponds to three linearly independent translations of the whole lattice. The latter show finite frequencies in the center of the 1BZ. Two rigid sublattices, containing cations and anions, respectively, are oscillating against each other. Thus for $k \equiv 0$,

$$u_{2n-1} = u_{2n+1} = u_{2n+3} = \dots \equiv u_{\oplus} \quad \text{for cations and}$$

$$u_{2n-2} = u_{2n} = u_{2n+2} = \dots \equiv u_{\ominus} \quad \text{for anions.}$$

Eqs.(3-1) and (3-2) then reduce to

$$m_{\ominus}\ddot{u}_{\ominus} = 2f(u_{\oplus} - u_{\ominus}) \quad , \tag{3-3}$$

$$m_{\oplus}\ddot{u}_{\oplus} = 2f(u_{\ominus} - u_{\oplus}) \quad . \tag{3-4}$$

These are simply the equations of two harmonic oscillators with finite eigenfrequencies. They correspond to those of the optical modes for $k \ll 10^7\ \text{cm}^{-1}$. The approximation made when replacing (3-1) and (3-2) by (3-3) and (3-4) eliminates the wave-vector dependence of the phonon frequencies in the region $10^7 \lesssim k \lesssim 10^8\ \text{cm}^{-1}$.

Frequency splittings of longitudinal (LO) and transverse (TO) optical phonons as described by the well-known Lyddane-Sachs-Teller relation cannot be derived from these equations.

From the simple electrostatic model illustrated in Fig.1 it can be seen, however, that in the direction of the phonon wave-vector of the LO mode there are electric fields of long range compared with the dimensions of the unit cell due to periodic concentrations of positive and negative charges. These electric fields induce additional restoring forces, leading to higher LO frequencies. The TO mode in Fig.1 on the other hand does not induce electric fields. The situation sketched here corresponds to the electrostatic approximation where the velocity of light is supposed to be infinite. If retardation effects are taken into account, every point in the vicinity of an atom will be displaced to the position occupied at time $t_0 - x/c$, where x is the distance and c the finite velocity of light. As a result, in a more realistic model, polar transverse phonons (with wave-vectors $k \lesssim 2\times10^4\ \text{cm}^{-1}$) will also be associated

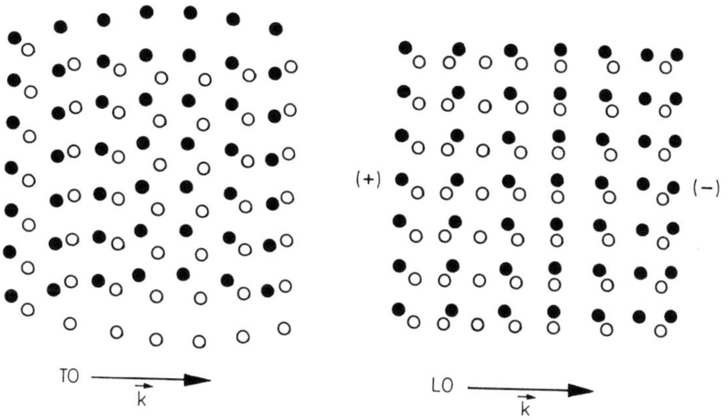

Fig.1 Transverse (TO) and longitudinal (LO) optical vibrations in a diatomic cubic crystal showing the long-range electrostatic fields that cause the LO-TO splitting. Sublattices with different charges are indicated by black and white circles (• = -; o = +), from /76/.

with electromagnetic fields having frequencies equal to those of the mechanical vibrations, see 3.2. If (3-3) and (3-4) are generalized, taking into account additional restoring forces caused by the local electric fields $E^{\ominus}_{loc}$ and $E^{\oplus}_{loc}$ at the positions of anions and cations, we get

$$m_{\ominus}\ddot{u}_{\ominus} = 2f(u_{\oplus} - u_{\ominus}) - e^{*}E^{\ominus}_{loc} \tag{3-5}$$

$$m_{\oplus}\ddot{u}_{\oplus} = 2f(u_{\ominus} - u_{\oplus}) + e^{*}E^{\oplus}_{loc} \quad , \tag{3-6}$$

where e^{*} denotes the effective charge of the ions. These two equations can be reduced to a single equation of motion with respect to the relative coordinate $(u_{\oplus} - u_{\ominus})$ by assuming the local fields $E^{\ominus}_{loc}$ and $E^{\oplus}_{loc}$ to be identical, i.e. $E^{\ominus}_{loc} = E^{\oplus}_{loc} \equiv E_{loc}$:

$$M(\ddot{u}_{\oplus} - \ddot{u}_{\ominus}) = -2f(u_{\oplus} - u_{\ominus}) + e^{*}E_{loc} \quad . \tag{3-7}$$

Herein $M = m_{\ominus} \cdot m_{\oplus}/(m_{\ominus} + m_{\oplus})$ is the reduced mass of an elementary cell.

In addition to this equation, we have to consider the electric polarization of the unit cell

$$P = (e^{*}/\upsilon_a)(u_{\oplus} - u_{\ominus}) + (\alpha_{\oplus} + \alpha_{\ominus})E_{loc}/\upsilon_a \quad . \tag{3-8}$$

$e^*(u_\oplus - u_\ominus)$ is a dipole caused by the displacement of the ions in the two sublattices and $\alpha_\oplus E_{loc}$ and $\alpha_\ominus E_{loc}$ are dipoles induced by the local electric fields. They deform the electron shells of the ions. $\alpha_\oplus$ and $\alpha_\ominus$ are the polarizabilities of cations and anions, respectively, and υ_a is the unit cell volume. For numerical calculations, we replace the local electric field in (3-7) and (3-8) by the macroscopic field $\underline{E}$ appearing in Maxwell's equations. This can be done by the well-known relation $\underline{E} = \underline{E}_{loc} - (4\pi/3)\underline{P}$ which, however, holds only for optically isotropic media /61/. If, finally, a generalized displacement coordinate defined by $\Omega = (M/\upsilon_a)^{1/2}\,(u_\oplus - u_\ominus)$ is introduced, (3-7) and (3-8) become of the form first claimed by Huang /36/ in 1951 for the description of cubic diatomic crystals with one polar mode

$$\ddot{\underline{Q}} = B^{11}\underline{Q} + B^{12}\underline{E} \qquad (3\text{-}9)$$

$$\underline{P} = B^{21}\underline{Q} + B^{22}\underline{E} \quad . \qquad (3\text{-}10)$$

The coefficients $B^{\mu\nu}$ can be interpreted macroscopically, see 3.3. The derivation given above allows us - at least for the simple cubic lattice in question - also to see directly the microscopic significance

$$B^{11} = -(2f/M) + (4\pi e^{*2}/3\upsilon_a M)\cdot\left[1-4\pi(\alpha_\oplus+\alpha_\ominus)/3\upsilon_a\right]^{-1} \quad , \qquad (3\text{-}11)$$

$$B^{12} = B^{21} = (e^*/\sqrt{M\upsilon_a})\left[1-4\pi(\alpha_\oplus+\alpha_\ominus)/3\upsilon_a\right]^{-1} \quad , \qquad (3\text{-}12)$$

$$B^{22} = \left[(\alpha_\oplus+\alpha_\ominus)/\upsilon_a\right]\cdot\left[1-4\pi(\alpha_\oplus+\alpha_\ominus)/3\upsilon_a\right]^{-1} \quad . \qquad (3\text{-}13)$$

$B^{12} = B^{21}$ trivially follows also from the derivation given in 3.4. In polyatomic crystals and others of lower symmetry the relation between the macroscopic and local electric fields cannot as a rule be expressed by a simple equation as used above. This makes the microscopic interpretation of the generalized coefficients $B^{\mu\nu}$ more complicated, see 4.13 and Appdx 2.

Because all directions in cubic crystals are optically equivalent, the $B^{\mu\nu}$ in (3-9) and (3-10) are scalars. Exactly longitudinal and transverse waves thus exist for all wave-vector directions in such materials.

If we combine (3-9) and (3-10) only with Maxwell's equation div $\underline{D} = 0$ for insulators, where $\underline{D} = \underline{E} + 4\pi\underline{P}$, we obtain a description of the long-wavelength optical phonons in the electrostatic approximation. For the electric field $\underline{E}$

$$\text{div}\,\underline{E} = -4\pi B^{21}(1-4\pi B^{22})^{-1} \cdot \text{div}\,\underline{Q} \quad . \tag{3-14}$$

Transverse waves (with lattice displacement vectors $\underline{Q}_T$) consequently do not induce electric fields because div $\underline{Q}_T = 0$. div $\underline{Q}_T = 0$ leads to either a vanishing or a homogeneous, frequency-independent electric field. Longitudinal waves on the other hand will be associated with a long-range macroscopic electric field because div $\underline{Q}_L \neq 0$. The generalized equation of motion (3-9) now includes a description of the frequency-splitting of polar TO and LO modes. Both TO and LO modes, however, still are without dispersion in the electrostatic approximation, i.e. their frequencies are wave-vector independent in the center of the 1BZ.

Additional Literature

Sindeev, Yu. G.: Determination of the effective charge of ions from the spectra of the Raman scattering of light by polaritons /240/.

3.2 THE RETARDATION EFFECT

The theory in the electrostatic approximation on the basis of Huang's equations (3-9) and (3-10) is developed by taking into account only div $\underline{D} = 0$. These equations determine the vectors $\underline{Q}$, $\underline{E}$ and $\underline{P}$. Regarding the total set of Maxwell's equations, the electrostatic treatment is equivalent to the approximation curl $\underline{E} = 0$. In a rigorous treatment this assumption has to be dropped. The rotation of the electric field then gives rise to a time-dependent magnetic field

$$\text{curl}\,\underline{E} = -(1/c)(\partial\underline{H}/\partial t) \tag{3-15}$$

and the magnetic field in turn causes an electric displacement

$$\text{curl}\ \underline{H} = (1/c)(\partial\underline{D}/\partial t) \quad , \tag{3-16}$$

with $\underline{D} = \underline{E} + 4\pi\underline{P}$. It can be seen directly from (3-15) that the approximation curl $\underline{E} = 0$ is identical to a model with an infinite velocity of light or with a time-independent magnetic field. Taking account of retardation effects for long-wavelength polar lattice vibrations thus means that Huang's equations are formally combined with the complete set of Maxwell's equations instead of with only div $\underline{D} = 0$.

It is well-known from crystal optics that Maxwell's equations (3-15) and (3-16) together with an plane-wave ansatz lead directly to a relationship between the electric field $\underline{E}$ and displacement $\underline{D}$ of the form

$$\underline{D} = n^2\left[\underline{E} - \underline{s}(\underline{s}\cdot\underline{E})\right] \quad , \tag{3-17}$$

where n is the refractive index of the medium. $\underline{s}$ denotes a unit vector in the direction of the wave-vector $\underline{k}$ and, because $\underline{s}$ is always perpendicular to $\underline{D}$ /61/, it follows from $\underline{D} = \underline{E} + 4\pi\underline{P}$ that

$$\underline{s}\cdot\underline{E} = -4\pi\underline{s}\cdot\underline{P} \tag{3-18}$$

and finally from (3-17) and (3-18)

$$\underline{E} = 4\pi(n^2-1)^{-1}\left[\underline{P} - n^2\underline{s}(\underline{s}\cdot\underline{P})\right] = 4\pi(n^2-1)^{-1}\,\underline{P}_T - 4\pi\underline{P}_L \quad . \tag{3-19}$$

Combining Huang's equations with (3-19) therefore leads to a theory which includes retardation effects. We now show that, in contrast to the electrostatic model, polar TO modes are here no longer without dispersion in the central region of the 1BZ.

We remember that the refractive index can always be written

$$n = ck/\omega \quad , \tag{3-20}$$

see (2-4). This follows directly from the definition $n = c/v$, where $v = \omega/k$ is the phase velocity of light in the medium. For transverse waves ($\underline{P} \perp \underline{s}$) the electric field according to (3-19) is simply given by

$$\underline{E} = 4\pi(n^2-1)^{-1} \underline{P}_T \quad , \tag{3-21}$$

whereas for purely longitudinal waves ($\underline{P} \parallel \underline{s}$)

$$\underline{E} = -4\pi\underline{P}_L \quad . \tag{3-22}$$

From (3-21) and (3-22) we can once more verify that the electrostatic approximation ($c \to \infty$), which according to (3-20) is identical with $n^2 \to \infty$, leads to vanishing electric fields for transverse modes, whereas the electric fields of longitudinal modes are not affected.

3.3 DISPERSION OF POLAR PHONON MODES IN THE POLARITON REGION

It has been shown that because of retardation effects polar transverse modes may also be associated with electric fields. This causes such modes to be coupled with light waves in the crystal when the energies $\hbar\omega$ and momenta $\hbar\underline{k}$ of the phonons and photons are approximately of the same order. In general the phonon frequencies of lattice waves lie in the infrared and the coupling with long-wavelength photons will thus take place in this region. Photons in the visible, on the other hand, propagate through the medium without resonances when their frequencies fall into the gap between IR and UV absorption (excitons).

Because of this coupling transverse polar modes will no longer be without dispersion in the center of the 1BZ. The dispersion relation can easily be derived from Huang's equations (3-9), (3-10), and (3-21) when making an ansatz of plane-wave solutions for $\underline{Q}$, $\underline{E}$ and $\underline{P} \sim \exp\{-i(\omega t-\underline{k}\cdot\underline{r})\}$.

The set of fundamental equations for TO modes is

$$-\omega^2\underline{Q} = B^{11}\underline{Q} + B^{12}\underline{E} \quad , \tag{3-23}$$

$$\underline{P} = B^{21}\underline{Q} + B^{22}\underline{E} \quad , \tag{3-24}$$

$$\underline{E} = 4\pi(n^2-1)^{-1} \underline{P} \quad . \tag{3-25}$$

Elimination of $\underline{Q}$, $\underline{E}$ and $\underline{P}$ leads to

$$n^2 = 1+4\pi B^{22} + 4\pi B^{12}B^{21}/(-B^{11}-\omega^2) \quad . \tag{3-26}$$

This is the well-known dispersion relation for the frequency-dependent dielectric function which is in general written macroscopically /61 §95, 62 §7/

$$n^2 = \varepsilon(\omega) = \varepsilon_\infty + \omega_T^2(\varepsilon_o-\varepsilon_\infty)/(\omega_T^2-\omega^2) \quad . \tag{3-27}$$

Damping has been neglected /61, 62/. ε_o denotes the 'static' dielectric constant for $\omega << \omega_T$ and ε_∞ the high-frequency dielectric constant for $\omega >> \omega_T$. ω_T is the frequency of the polar transverse lattice wave in the electrostatic approximation. $\varepsilon_o-\varepsilon_\infty \equiv 4\pi\rho$ denotes the oscillator strength of the mode. In practice, ε_∞ has to be determined at some frequency in the gap between IR and UV absorption, which for transparent materials is the visible region (e.g. the laser frequency), and ε_o at some frequency that is small compared with that of the optical lattice vibration but large compared with the acoustic phonon frequencies. This means measurements at $\sim 10^2$ MHz.

Comparing (3-26) and (3-27), we can now see the macroscopic interpretations of the coefficients $B^{\mu\nu}$, which are of great importance experimentally:

$$B^{11} = -\omega_T^2 \quad , \tag{3-28}$$

$$B^{12} = B^{21} = \omega_T\left[(\varepsilon_o-\varepsilon_\infty)/4\pi\right]^{1/2} = \omega_T\sqrt{\rho} \quad , \tag{3-29}$$

$$B^{22} = (\varepsilon_\infty-1)/4\pi \quad . \tag{3-30}$$

Taking (3-20) into account, the dispersion relation for <u>transverse polar phonons</u> according to (3-27) becomes

$$(c^2k^2/\omega^2) = \varepsilon_\infty + \omega_T^2(\varepsilon_o-\varepsilon_\infty)/(\omega_T^2-\omega^2) \quad . \tag{3-31}$$

As this equation is quadratic in ω^2, there are two solutions $\omega^2 = \omega^2(k)$, corresponding to the two TO branches sketched in Fig.2a.

In addition to these transverse branches there is a <u>longitudinal branch.</u> The corresponding solution is obtained when combining

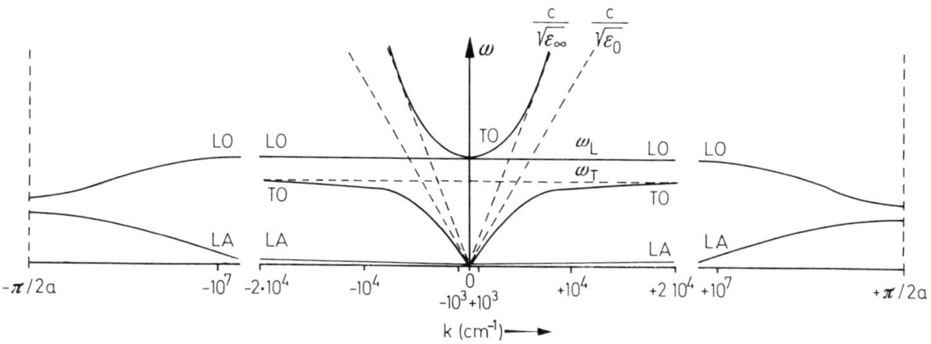

Fig.2a Dispersion curves of the long-wavelength optical phonons, photons and polaritons of a cubic diatomic lattice in the center of the 1BZ. In order to demonstrate the connection with dispersion effects in the region $10^6 \lesssim k \lesssim 10^8$ cm^{-1}, the branches of a LO and a LA phonon have been added in a different linear scale, from /76/.

Huang's equations (3-23) and (3-24) with (3-22) instead of (3-21). Elimination of $\underline{Q}$, $\underline{E}$ and $\underline{P}$ then leads to the Lyddane-Sachs-Teller relation:

$$\omega^2 = 4\pi B^{12} B^{21} (1+4\pi B^{22})^{-1} - B^{11} \tag{3-32}$$

which due to (3-28), (3-29) and (3-30) can be written in the well-known form

$$\omega_L^2/\omega_T^2 = \varepsilon_o/\varepsilon_\infty \quad . \tag{3-33}$$

$\omega = \omega_L$ thus does not depend on k in the polariton region and accordingly the LO branch in Fig.2a is without dispersion.

The detailed interpretation of Fig.2a is as follows: In the framing pictures to left and right a longitudinal optical (LO) and longitudinal acoustic (LA) dispersion branch as derived from lattice dynamics on the basis of the Born-von Kármán model have been sketched for $10^7 \lesssim k \lesssim 10^8$ cm^{-1}. Dispersion effects caused by the coupling between electromagnetic waves and transverse polar phonons appear in the center of the 1BZ for $10^3 \lesssim k \lesssim 2 \times 10^4$ cm^{-1}. This 'polariton region' has been sketched on a different linear scale (middle picture). The dispersion curve of photons in the absence of polar lattice waves is indicated by a dashed line with slope $c/\sqrt{\varepsilon_\infty}$. Due to the coupling with lattice waves, the photons acquire some phonon character for decreasing k and the dispersion curve ends at the frequency of the LO phonon. On the other hand, the lowest-frequency TO phonon acquires some photon character for decreasing k, and the dispersion branch reaches $\omega = 0$ for $k = 0$ with the slope $c/\sqrt{\varepsilon_0}$ in the origin. This can easily be derived from (3-31). The LO branch, (3-33), is finally described by a horizontal line without dispersion. The energies of the quanta of the two TO branches in the polariton region are partly mechanical and partly electromagnetic /36, 62/, see 4.7, 4.11 and Fig.2b. Polaritons associated with the upper TO branch are referred to here as 'photon-like' and those originating from the lower TO branch as 'phonon-like'.

For $k = 0$ the upper TO branch has the same frequency as the LO phonon, which implies that the electric field due to the retardation effect is identical to that of the LO phonon. This can be seen directly from (3-21) and (3-22) because $n^2 = 0$ holds for $k = 0$, see again (3-20). Two rigid sublattices are oscillating against each other and there are obviously no wave-vector directions in the center of the 1BZ. Hence, LO and TO modes can no longer be distinguished from each other.

In the electrostatic approximation ($c \to \infty$), the slopes of the two dashed lines $c/\sqrt{\varepsilon_\infty}$ and $c/\sqrt{\varepsilon_0}$ in Fig.2a would be infinite and all the polariton dispersion effects would coincide with the ω axis. The lowest-frequency polar TO branch then becomes the horizontal dashed line at $\omega = \omega_T$. Furthermore, no photon dispersion curves would appear in the ω, k diagram. Note, however, that the LO-TO splitting described by (3-33) still takes place.

According to (3-19) and (3-20) the electrostatic approximation ($n^2 \to \infty$) can be achieved for $k \to \infty$ in the polariton theory. Since this theory, however, is an approximation valid only for long-wavelength lattice waves, $\omega = \omega(k)$ for $k \to \infty$ describes only phonons existing in the region $10^5 \lesssim k \lesssim 10^7$ cm^{-1} of the 1BZ. The phase velocities of these phonons vanish. The phase velocity of photons in the absence of polar phonons on the other hand would be $c/\sqrt{\varepsilon_\infty}$ (slope of the light line). Hence, the phase velocities of the mixed excitations described by the phonon-like polariton branch change from 0 to $c/\sqrt{\varepsilon_0}$ for decreasing k values (the respective slopes of the dispersion curve).

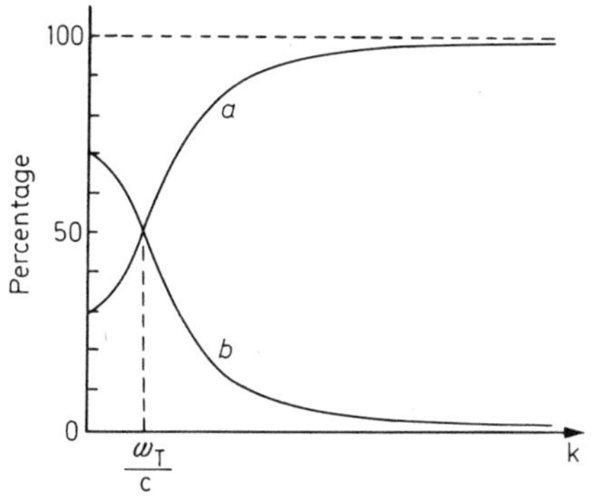

Fig.2b Percentage of mechanical energy in the (a) phonon- and (b) photon-like transverse branches, from /36/.

The horizontal parts of the dispersion curves of the polar TO phonons in Fig.2a obviously describe lattice waves with different wavelengths but identical frequencies. At the short wavelength limit of the 1BZ, the oscillating dipole moments of neighboring elementary cells are cancelled so that no macroscopic electric fields are built up. Such fields are created by the lattice waves only if their wavelengths become large enough, i.e. of the order of uncoupled electromagnetic waves in the medium. This is exactly what happens in the polariton region.

Polaritons are excitations in crystalline media, which always exist because the materials are in thermal contact with their surroundings: they are not excited only by external electric fields entering the medium. When, on the other hand, an infrared electromagnetic wave with a frequency lying in the polariton region (ordinate in Fig.2) enters the medium,it travels through it as a polariton wave with a wave vector k' which can easily be found on the abscissa in the ω, k diagram. The phase velocity of this wave is $(\partial\omega/\partial k)_{k'}$. Finally,polaritons may also be created in Stokes-Raman scattering processes, as described by (1-1). In this case only part of the incident photon energy $\hbar\omega_i$ traverses the medium as a polariton wave $\hbar\omega$, and the rest continues at much higher frequency as a pure photon $\hbar\omega_s$ with the phase velocity $c/\sqrt{\varepsilon_\infty}$. As can be seen from the Boltzmann factor,only a small part of the polariton population at room temperature is created in this way, most of them being created thermally. The phenomenon is beautifully illustrated by Anti-Stokes scattering from polaritons, see /284, Figs.1 and 2/. Polaritons which have previously been excited thermally are here destroyed and add their energies to the incident photons, (1-1). The decreasing intensity in the Anti-Stokes spectra for decreasing temperatures shows the decreasing thermal population at lower temperatures.

Huang's model may in summary be characterized in the following way. The wave-vector dependence of phonon frequencies due to interatomic forces has been eliminated by introducing k-independent coefficients $B^{\mu\nu}$ and by the statements leading to (3-3) and (3-4), where the lattice displacement coordinate $(u_{\oplus} - u_{\ominus}) \sim Q$ is first introduced. On the other hand, $\underline{k}$ dependence is reintroduced later by plane waves in combination with Maxwell's equations. For this reason, the theory describes polariton dispersion, but not dispersion effects in the region $10^7 \lesssim k \lesssim 10^8\ cm^{-1}$.

3.4 ENERGY DENSITY AND POLARITONS IN MAGNETIC MATERIALS

We now turn our attention to the energy in a material with polariton dispersion. As first shown by Huang /36/, the following expression correctly describes the energy density (Hamilton density) in such a medium

$$H = (1/2)\dot{\underline{Q}}^2 - (1/2)\underline{Q}B^{11}\underline{Q} - \underline{Q}B^{12}\underline{E} - (1/2)\underline{E}B^{22}\underline{E} + \underline{E}\cdot\underline{P} + (1/8\pi)(\underline{E}^2+\underline{H}^2) . \quad (3\text{-}34)$$

Energy conservation requires that $\partial H/\partial t$ remains zero. From (3-34) we therefore derive

$$\partial H/\partial t = \dot{\underline{Q}}(\ddot{\underline{Q}}-B^{11}\underline{Q}-B^{12}\underline{E}) + \dot{\underline{E}}(\underline{P}-B^{12}\underline{Q}-B^{22}\underline{E}) + \underline{E}\cdot\dot{\underline{P}} + (1/4\pi)(\underline{E}\cdot\dot{\underline{E}}+\underline{H}\cdot\dot{\underline{H}}) \quad . \quad (3\text{-}35)$$

The first two terms obviously vanish because Huang's equations (3-9) and (3-10) hold. The remaining terms can be rewritten

$$(1/4\pi)\left[\underline{E}(\dot{\underline{E}}+4\pi\dot{\underline{P}})+\underline{H}\cdot\dot{\underline{H}}\right] = (1/4\pi)(\varepsilon\ \underline{E}\cdot\dot{\underline{E}}+\underline{H}\cdot\dot{\underline{H}}) \quad . \quad (3\text{-}36)$$

Eq.(3-36), however, is the time derivative of

$$(1/8\pi)(\varepsilon\ \underline{E}^2+\underline{H}^2) \quad (3\text{-}37)$$

representing the energy density of the electromagnetic field in a nonmagnetic material. From thermodynamic considerations it is well-known that the equation of motion as well as the macroscopic polarization can be derived from the potential energy density when the contributions of the vacuum field are omitted:

$$\ddot{\underline{Q}} = -\partial\Phi/\partial\underline{Q} \tag{3-38}$$

and

$$\underline{P} = -\partial\Phi/\partial\underline{E} \quad , \tag{3-39}$$

see /304, 138/. In order to apply (3-38) and (3-39), the energy density H (3-34) therefore has to be reduced by the first term $(1/2)\dot{\underline{Q}}^2$ describing the kinetic energy and by the last three terms. $(1/4\pi)(\underline{E}^2+\underline{H}^2)$ obviously represents the vacuum energy density and $\underline{E}\cdot\underline{P}$ the potential energy of all dipoles in the material (caused by lattice vibrations) in an external field $\underline{E}$. Eq.(3-34) reduces to

$$\Phi = -(1/2)\underline{Q}B^{11}\underline{Q} - \underline{Q}B^{12}\underline{E}-(1/2)\underline{E}B^{22}\underline{E} \quad . \tag{3-40}$$

As can easily be seen, Huang's equations (3-9) and (3-10) are in fact immediately derived from (3-40) when (3-38) and (3-39) are applied, and it trivially holds that $B^{12} = B^{21}$.

The percentage of the mechanical energy of the phonon- and photon-like dispersion branches as a function of the wave vector has been calculated by Huang /36/ from (3-34), see Fig.2b.

The generalization of (3-40) to include materials with a magnetic permeability μ different from 1 becomes

$$\Phi = -(1/2)\underline{Q}B^{11}\underline{Q}-\underline{Q}B_E^{12}\underline{E}-\underline{Q}B_H^{12}\underline{H}-(1/2)\underline{E}B_E^{22}\underline{E}-(1/2)\underline{H}B_H^{22}\underline{H} \quad . \tag{3-41}$$

The two additional terms describe the interaction of lattice vibrations with the magnetic field $\underline{H}$ and the potential energy of magnetic dipoles (induced by the $\underline{H}$ field) in the $\underline{H}$ field.
Direct interactions between the electric and magnetic fields in the material have been left out of consideration in this approximation, i.e. a mixed term in $\underline{E}$ and $\underline{H}$ has been neglected. The third fundamental equation (3-17) or (3-19) now has to be replaced by two relations which are similarly derived from Maxwell's equations by

taking into account that $\mu \neq 1$. The total set of fundamental equations then becomes

$$\underline{P} = (c^2k^2/\omega^2)\left[(\varepsilon-1)/4\pi\varepsilon\mu\right]\left[\underline{E}-\underline{s}(\underline{s}\cdot\underline{E})\right] \quad , \tag{3-41}$$

$$\underline{M} = (c^2k^2/\omega^2)\left[(\mu-1)/4\pi\varepsilon\mu\right]\left[\underline{H}-\underline{s}(\underline{s}\cdot\underline{H})\right] \quad , \tag{3-42}$$

$$\ddot{\underline{Q}} = -(\partial\Phi/\partial\underline{Q}) = B^{11}\underline{Q} + B^{12}_{E}\underline{E} + B^{12}_{H}\underline{H} \quad , \tag{3-43}$$

$$\underline{P} = -(\partial\Phi/\partial\underline{E}) = B^{12}_{E}\underline{Q} + B^{22}_{E}\underline{E} \quad , \tag{3-44}$$

$$\underline{M} = -(\partial\Phi/\partial\underline{H}) = B^{12}_{H}\underline{Q} + B^{22}_{H}\underline{H} \quad . \tag{3-45}$$

$\underline{M} = \left[(\mu-1)/4\pi\right]\underline{H}$ denotes the magnetic polarization. The system (3-41) to (3-45) is linear in $\underline{Q}$, $\underline{E}$, $\underline{P}$, $\underline{H}$ and $\underline{M}$ and can be solved by the well-known methods of linear algebra. We do not further discuss magnetic polaritons because no experimental work on the subject has yet been published. It should, however, be mentioned that the results suggest that the dispersion curves are of a similar shape to those shown in Fig.2a. Mills and Burstein /300/ have presented dispersion branches of magnetic polaritons in MnF_2 for a vanishing external magnetic field. We refer to this article for further discussion.

4. Dispersion of Polar Optical Modes in Polyatomic General Crystals

4.1 FUNDAMENTAL EQUATIONS OF THE POLARITON THEORY

In 3.1 to 3.4 we have presented a detailed discussion of the physical background of the polariton theory on the basis of its simplest model. We now present a more rigorous derivation of the general theory for arbitrary crystals, as developed over the last few years /31, 63 to 68, 73, 332/. Furthermore, we shall discuss experimental methods and results demonstrating the most characteristic phenomena /31, 35, 69 to 78/.

Following Cochran and Cowley /79/, the two fundamental equations describing the long-wavelength lattice waves in a polyatomic crystal may be derived from a potential energy density Φ which in the harmonic approximation can be written

$$\Phi = (1/2)\underline{u}\cdot L\cdot\underline{u} - \underline{u}\cdot M\cdot\underline{E} - (1/2)\underline{E}\cdot N\cdot\underline{E} \quad . \tag{4-1}$$

$\underline{u}$ is a 3ℓ dimensional displacement vector for the ℓ atoms in the unit cell. L is a $(3\ell \times 3\ell)$ matrix, M a $(3\ell \times 3)$ matrix, and N a (3×3) matrix. The relation obviously represents the generalization of (3-40) to arbitrary crystals. Because only infrared-active modes show polariton dispersion,we can exclude the 3p displacement-vector components of silent modes, only Raman-active modes, and the three nonpolar acoustical mode components. All components of the corresponding 3p + 3 lines in the matrix M are O. $\underline{u}$ reduces to become of dimension $3n = 3\ell - 3p - 3$, and the matrices L, M, and N become of type $(3n \times 3n)$, $(3n \times 3)$ and (3×3), respectively. The equation of motion and the macroscopic polarization can again be derived from (4-1) as

$$\rho\ddot{\underline{u}} = -\partial\Phi/\partial\underline{u} \quad , \tag{4-2}$$

and

$$\underline{P} = -\partial\Phi/\partial\underline{E} \quad . \tag{4-3}$$

In (4-2) ρ is a diagonal matrix with elements $\rho_i = m_i/\upsilon_a (i = 1...3n)$. The m_i are atomic masses and υ_a is the cell volume. Groups of three elements of the ρ_i are always identical and represent the densities of the n sublattices, appearing threefold because there are three degrees of freedom. These densities can be eliminated from the fundamental equations by defining a generalized displacement vector $\underline{u}'$ with components $u_i' = \sqrt{\rho_i}\, u_i$ and by introducing corresponding matrices with elements $L_{ij}' = L_{ij}(\rho_i\rho_j)^{-1/2}$ and $M_{i\alpha}' = M_{i\alpha}(\rho_i)^{-1/2}$. In analogy to the normal coordinate treatment in molecular physics, it is useful to diagonalize L' when describing the purely mechanical part of the energy density: $-A^{-1}\cdot L'\cdot A = B''$ where $A^{-1} = A^{+}$. In general, however, the second mixed term in (4-1) makes an important contribution to the energy density Φ. The linear orthogonal transformation applied thus does not define normal coordinates in the usual sense. The 'normal coordinates' obtained in general are not linearly independent of each other but are still coupled due to the electric field. We therefore use the term 'quasinormal coordinates'. These are $\underline{Q} = A^{-1}\cdot\underline{u}'$ and the remaining two terms of Φ become $B^{12} = A^{-1}\cdot M'$ and $B^{22} \equiv N$. Eq.(4-1) can be written alternatively as

$$\Phi = -(1/2)\underline{Q}\cdot B^{11}\cdot\underline{Q} - \underline{Q}\cdot B^{12}\cdot\underline{E} - (1/2)\underline{E}\cdot B^{22}\cdot\underline{E} \quad . \tag{4-4}$$

Taking into account (4-2) and (4-3), the equation of motion and the macroscopic polarization are derived as

$$\ddot{\underline{Q}} = -(\partial\Phi/\partial\underline{Q}) = B^{11}\cdot\underline{Q} + B^{12}\cdot\underline{E} \quad , \tag{4-5}$$

$$\underline{P} = -(\partial\Phi/\partial\underline{E}) = B^{21}\cdot\underline{Q} + B^{22}\cdot\underline{E} \quad , \tag{4-6}$$

where B^{21} is the transposed matrix of B^{12}, i.e. $B^{21} = (B^{12})^{+}$. With plane-wave solutions for $\underline{Q}$, $\underline{E}$ and $\underline{P} \sim \exp[-i(\omega t-\underline{k}\cdot\underline{r})]$, (4-5), (4-6), and (3-19) represent the fundamental set of equations of the polariton theory for general crystals:

$$-\omega^2 \underline{Q} = B^{11} \cdot \underline{Q} + B^{12} \cdot \underline{E} \quad , \tag{4-7}$$

$$\underline{P} = B^{21} \cdot \underline{Q} + B^{22} \cdot \underline{E} \quad , \tag{4-8}$$

$$\underline{E} = 4\pi(n^2-1)^{-1}[\underline{P} - n^2 \underline{s}(\underline{s} \cdot \underline{P})] \quad . \tag{4-9}$$

The quasinormal coordinates are normal coordinates in the usual sense only for vanishing electric fields. This happens for $\underline{k} \to \infty$ and wave-vectors in the principal crystal directions where there are exactly transverse polar phonons, see 3.2. For silent modes and only Raman-active modes, on the other hand, the normal coordinate treatment still holds in the classic form because these modes are never associated with electric fields for any wave-vector directions and magnitudes. Only acoustic modes in piezoelectric crystals can be associated with electric fields due to the piezoelectric effect. This may cause some polariton dispersion on the corresponding branches.

Additional Literature

Inomata, H., Horie, C.: The master equation in a polariton system /256/.

4.2 THE MACROSCOPIC THEORY AND FRESNEL'S EQUATION OF THE WAVE NORMAL

Microscopic interpretations of the generalized coefficients $B^{\mu\nu}$ in (4-7) and (4-8) have been discussed in /263/ and /80/, see 4.13 and Appdx 2. The coefficients may, however, easily be related also to macroscopic properties that can be experimentally determined. If we restrict our discussion to cubic, uniaxial, and orthorhombic crystals, the components of (4-7) and (4-8) for the principal crystallographic axes (α = 1, 2 or 3) are

$$-\omega^2 Q_{\alpha i} = B^{11}_{\alpha ii} Q_{\alpha i} + B^{12}_{\alpha i} E_\alpha \quad , \tag{4-10}$$

and

$$P_\alpha = \sum_{i=1}^{n_\alpha} B^{12}_{\alpha i} Q_{\alpha i} + B^{22}_{\alpha} E_\alpha \quad . \tag{4-11}$$

n_α is the number of polar modes for the direction α and $n_1+n_2+n_3 = n$ the total number of polar modes. If the symmetry of the crystal is at least orthorhombic, the quasinormal coordinates $Q_{\alpha i}$ for the α direction are linear combinations of only those components of $\underline{E}$ and $\underline{P}$ that refer to the same axis, α. Every quasinormal coordinate describes a mode with lattice displacements parallel to one of the three principal directions. Consequently, there are at least two TO and one LO mode for wave-vectors along every principal direction.

In monoclinic and triclinic systems the crystallographic axes are no longer orthogonal. Pure TO and LO modes therefore generally exist only for wave-vectors in certain off-principal directions. This is because the axes of the dielectric tensor ellipsoid are no longer determined by the crystallographic axes. Phonons propagating in arbitrary directions, on the other hand, are in general of mixed type due to multimode mixing. Monoclinic and triclinic crystals are further discussed in 4.9.

The macroscopic theory has so far been worked out in detail only for the orthorhombic system and such of higher symmetries. This is partly because the experimental work that must be done to assign all fundamental LO and TO modes in monoclinic and triclinic systems is extremely extensive and the authors do not know of its existence for even one material. The difficulties may be appreciated by reference to 4.5, 4.6, and 4.7, where a rather simple trigonal crystal is described.

It is convenient to study three different relations for the frequency-dependent dielectric function $\varepsilon = \varepsilon(\omega)$. The first can easily be derived from (4-10) and (4-11) by eliminating $\underline{Q}_{\alpha i}$ and taking into account that $\varepsilon(\omega)\underline{E} = \underline{E} + 4\pi\underline{P}$. The second is a well-known relation from optical dispersion theory, where $S_i \equiv 4\pi\rho_i$ are the oscillator strengths of the infrared active modes. The third is a factorized form derived by Kurosawa /81/.

$$\varepsilon_\alpha(\omega) = 1 + 4\pi B^{22}_{\alpha\alpha} + \sum_{i=1}^{n_\alpha} \frac{4\pi (B^{12}_{\alpha i})^2}{-B^{11}_{\alpha ii} - \omega^2} , \qquad (4\text{-}12)$$

$$\varepsilon_\alpha(\omega) = \varepsilon_{\alpha\infty} + \sum_{i=1}^{n_\alpha} \frac{S_i \omega^2_{\alpha Ti}}{\omega^2_{\alpha Ti} - \omega^2} , \qquad (4\text{-}13)$$

$$\varepsilon_\alpha(\omega) = \varepsilon_{\alpha\infty} \prod_{i=1}^{n_\alpha} \frac{\omega^2_{\alpha Li} - \omega^2}{\omega^2_{\alpha Ti} - \omega^2} \quad . \tag{4-14}$$

Fig.3 is a graphical presentation of the Kurosawa relation (4-14). For $\omega \to 0$, (4-14) reduces to the well-known generalized LST relation:

$$(\varepsilon_\alpha(0)/\varepsilon_{\alpha\infty}) = \prod_{i=1}^{n_\alpha} (\omega^2_{\alpha Li}/\omega^2_{\alpha Ti}) \quad . \tag{4-15}$$

$\varepsilon_\alpha(0) \equiv \varepsilon_{\alpha 0}$ denotes the static dielectric constant. For $\omega \to \infty$, on the other hand, the product in (4-14) becomes 1 and $\varepsilon_\alpha = \varepsilon_{\alpha\infty}$. Because damping terms have been neglected in the resonance denominators, the frequencies of the n_α TO modes are described mathematically by the poles appearing for $\omega \to \omega_{\alpha Ti}$. The frequencies of the LO modes are obtained in the same way from the zeros of the dielectric function where $\omega = \omega_{\alpha Li}$. From the resonance denominators in (4-12) and (4-13) it can be seen that $B^{11}_{\alpha ii} = -\omega^2_{\alpha Ti}$, which are the negative squares of the frequencies of transverse phonons in α direction. For $\omega \to \infty$, on the other hand, it follows from (4-12) and (4-13) that $B^{22}_{\alpha\alpha} = (\varepsilon_\infty - 1)/4\pi$. Finally, (4-12) and (4-14) determine

$$(B^{12}_{\alpha k})^2 = (\varepsilon_{\alpha\infty}/4\pi) \frac{\prod_{j=1}^{n_\alpha} (\omega^2_{\alpha Lj} - \omega^2_{\alpha Tk})}{\prod_{\substack{j=1 \\ j \neq k}}^{n_\alpha} (\omega^2_{\alpha Tj} - \omega^2_{\alpha Tk})} \quad , \tag{4-16}$$

when the constants $B^{11}_{\alpha ii}$ and $B^{22}_{\alpha\alpha}$ are explicitly substituted and $\omega = \omega_{\alpha Tk}$, which is the frequency of the k-th transverse mode in α direction. Hence, the coefficients in (4-10) and (4-11) can be experimentally determined when the frequencies $\omega_{\alpha Tj}$ (TO modes) and $\omega_{\alpha Lj}$ (LO modes) and the dielectric constants $\varepsilon_{\alpha\infty}$ are known. The mode strength S_k of the k-th mode can be derived from (4-12) and (4-13) for $\omega \to \omega_{\alpha Tk}$

$$S_k = (4\pi/\omega^2_{\alpha Tk})(B^{12}_{\alpha k})^2 \quad . \tag{4-17}$$

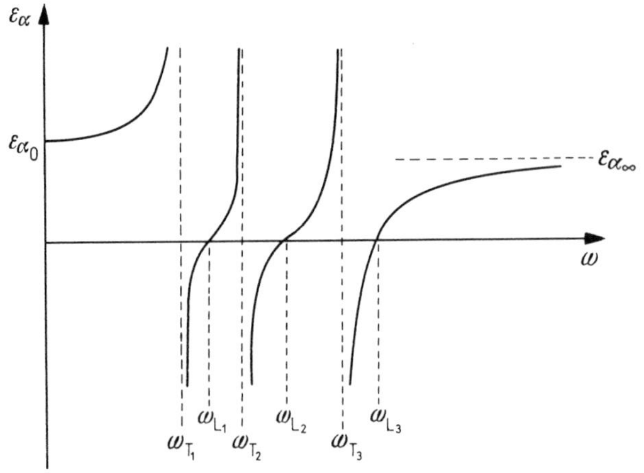

Fig.3 The dielectric function $\varepsilon = \varepsilon(\omega)$ as determined by the Kurosawa relation when neglecting damping (4-14). The graphs are sketched for one principal direction, α = 1,2 or 3, and for n_α = 3 polar modes.

Substituting (4-16) in (4-17), we obtain

$$S_k = (\omega^2_{\alpha Lk} - \omega^2_{\alpha Tk})(\varepsilon_{\alpha\infty}/\omega^2_{\alpha Tk}) \prod_{\substack{j=1 \\ j\neq k}}^{n_\alpha} \frac{(\omega^2_{\alpha Lj} - \omega^2_{\alpha Tk})}{(\omega^2_{\alpha Tj} - \omega^2_{\alpha Tk})} \quad . \tag{4-18}$$

The determination of the LO frequencies $\omega_{\alpha Lj}$ for calculations of the coefficients $B^{\mu\nu}$ can thus be replaced by the determination of the mode strengths S_k. For this purpose IR-intensity measurements are required. According to (4-18) the mode strength is proportional to the magnitude of the LO-TO splitting of the corresponding mode (first factor). S_k thus vanishes for $\omega_{\alpha Lk} = \omega_{\alpha Tk}$.

In order to solve the set (4-9), (4-10), and (4-11), it is convenient to combine the vectors $\underline{Q}$, $\underline{E}$ and $\underline{P}$ to a (n + 3 + 3)-dimensional vector $\underline{X} = (\underline{Q}, \underline{E}, \underline{P})$. The equations can then be written in the comprehensive form

$$A \cdot \underline{X} = 0 \quad . \tag{4-19}$$

The matrix A is of dimension (n + 6) × (n + 6)

$$A = \begin{pmatrix} B^{11} + \omega^2 I & B^{12} & O \\ (B^{12})^+ & B^{22} & -I \\ O & -I & T \end{pmatrix} \quad .$$

I denotes the three-dimensional unit matrix, O a zero-matrix, and T stands for $4\pi(n^2-1)^{-1}\cdot(I-n^2\underline{s}\,\underline{s})$. (The refractive index n in this relation must not be confused with the total number of polar modes $n_1 + n_2 + n_3 = n$, see above.) The condition for a nontrivial solution of (4-19) is: det A = O. This leads to a generalized Fresnel equation

$$\frac{s_1^2}{\frac{1}{n^2} - \frac{1}{\varepsilon_1(\omega)}} + \frac{s_2^2}{\frac{1}{n^2} - \frac{1}{\varepsilon_2(\omega)}} + \frac{s_3^2}{\frac{1}{n^2} - \frac{1}{\varepsilon_3(\omega)}} = 0 \quad , \tag{4-20}$$

where $\underline{s} = (s_1,s_2,s_3)$ is a unit vector in the direction of the wave vector $\underline{s} = \underline{k}/|\underline{k}|$, see /66/. If we compare (4-20) with the corresponding relation from crystal optics, the only difference is that the dielectric functions $\varepsilon_\alpha(\omega)$ are frequency-dependent and not simply constants valid for a certain frequency. The frequency dependence is given explicitly by the Kurosawa relation (4-14) or by (4-13). Written without denominators, (4-20) becomes

$$\sum_{\alpha=1}^{3} \varepsilon_\alpha(\omega)\; s_\alpha^2 \prod_{\substack{\beta=1 \\ \beta\neq\alpha}}^{3} (\varepsilon_\beta(\omega) - n^2) = 0 \quad . \tag{4-21}$$

So as not to exclude any solutions, the equation will be discussed in this form. Taking into account the relation $n^2 = c^2k^2/\omega^2$ (3-20), (4-21) represents the general dispersion relation of polaritons in orthorhombic crystals and such of higher symmetries.

The k dependence of the mode frequencies was introduced by plane-wave solutions in (4-7) and (4-8), whereas the optical anisotropy was allowed for by introducing (4-9), where the wave normal vector $\underline{s}$ first appears (crystal optics). Eq.(4-21) is of power (n + 2) in ω^2 and there are (n + 2) polariton eigenfrequencies for every direction of the wave-vector. Directional dispersion as well as polariton dispersion in matter is completely described by (4-21) and we are now going to discuss some special cases concerning crystal classes which have been experimentally studied.

Additional Literature

Barentzen, H., Schrader, B., Merten, L.: Optical properties of lattice vibrations in molecular crystals /228/.

Tsu, R., Iha, S.S.: Phonon and polariton modes in a superlattice /229/.

4.3 POLARITONS IN CUBIC CRYSTALS

In cubic crystals $\varepsilon_1(\omega) = \varepsilon_2(\omega) = \varepsilon_3(\omega) \equiv \varepsilon(\omega)$ because they are optically isotropic. Accordingly, the numbers of eigenfrequencies for all (principal) directions are equal: $n_1 = n_2 = n_3 \equiv m$. Eq.(4-21) becomes

$$\varepsilon(\omega)\left[\omega^2\varepsilon(\omega) - c^2k^2\right]\sum_{\alpha=1}^{3} s_\alpha^2 = 0 \tag{4-22}$$

with (3-20), and because $s_1^2 + s_2^2 + s_3^2 = 1$, it follows that

$$\varepsilon(\omega)\left[\omega^2\varepsilon(\omega) - c^2k^2\right] = 0 \quad . \tag{4-23}$$

This equation has two different types of directionally independent solutions:

a) $\varepsilon(\omega) = 0$ leads to $\omega = \omega_{Li}$, the frequencies of the LO phonons; see the discussion of the Kurosawa relation in 4.2. Their 'dispersion relation' is the generalized LST relation and the LO branches thus do not depend on the magnitude of the wave-vector $\underline{k}$ within the whole polariton region.

b) $\omega^2\varepsilon(\omega) - c^2k^2 = 0$ or $\varepsilon(\omega) = c^2k^2/\omega^2 = n^2$ describes the wave vector-dependent polariton-dispersion branches. All polariton frequencies in cubic crystals are directionally independent since only k^2 appears in this relation. Note, however, that the Raman-scattering intensities in cubic crystals are direction-dependent, see 4.12. Thus, for experimental purposes, all $\underline{k}$ directions are not equivalent. For $k \to 0$, the frequencies are identical to those of the LO phonons, either $\omega = \omega_{Li}$ or $\omega = 0$. For $k \to \infty$, on the other hand, ω can be either $\omega = \omega_{Ti}$ or $\omega \to \infty$. This can be seen from (4-14), written

$$(1/\omega^2) \prod_{i=1}^{m} \frac{\omega_{Ti}^2 - \omega^2}{\omega_{Li}^2 - \omega^2} = \varepsilon_\infty/c^2k^2 \quad , \tag{4-24}$$

where $\varepsilon(\omega) = c^2k^2/\omega^2$ has been introduced.

For the simplest type of diatomic cubic crystal with only two atoms in the unit cell, discussed in detail in 3.3, m = 1 holds for every direction. Consequently, the vector of the quasinormal coordinates has only three components:

$$\underline{Q} = \begin{pmatrix} Q_1 \\ Q_2 \\ Q_3 \end{pmatrix} \quad .$$

Eq.(3-17) and $\underline{D} = \varepsilon(\omega)\cdot\underline{E}$ furthermore require $\underline{s}(\underline{s}\cdot\underline{E}) = 0$. This implies $\underline{s}\perp\underline{E}$ for nonvanishing wave vectors and electric fields. From Huang's equation (3-23) it can easily be realized that polaritons in cubic crystals are exactly transverse ($\underline{s}\perp\underline{Q}$) because

$$(s_1,s_2,s_3)\begin{pmatrix} Q_1 \\ Q_2 \\ Q_3 \end{pmatrix}(-\omega^2-B^{11}) = B^{12}(s_1,s_2,s_3)\begin{pmatrix} E_1 \\ E_2 \\ E_3 \end{pmatrix} = 0 \quad . \tag{4-25}$$

The macroscopic polarization $\underline{P}$ also stands perpendicular to $\underline{s}$, see (3-18).

For all wave-vector directions there are purely longitudinal and transverse modes. This result is unique for cubic crystals and does not apply to crystals of lower symmetry. The proof can easily be extended to cubic crystals with an arbitrary number of polar modes by arranging the quasinormal coordinates in groups of three, each group referring to a certain mode. Fig.4 shows the polariton-dispersion curves in a cubic crystal with m = 2 polar modes. The second (quadratic) factor in (4-23) describing the transverse polaritons determines three dispersion branches because $\omega^2\cdot\varepsilon(\omega)-c^2k^2=0$ is of power m + 1 in ω^2. Every branch becomes doubly degenerate for all wave-vector directions, but this has no physical significance. See, however, the discussion of uniaxial and biaxial crystals in 4.4 and 4.9. The lowest-frequency dispersion branch goes to zero energies for $k \to 0$ and the next one correspondingly couples down from ω_{T2} to ω_{L1}. The highest-frequency one finally ends at ω_{L2}

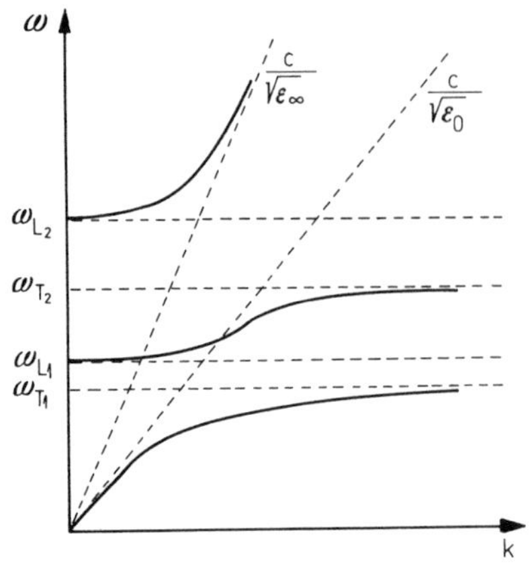

Fig.4 Polariton-dispersion branches in a cubic crystal with two polar modes see /160,166/.

in the center of the 1BZ. In the regions between ω_{T1} and ω_{L1} and ω_{T2} and ω_{L2}, respectively, there are no bulk polaritons. We shall later see that these are the energy regions where surface polaritons can be excited. For $k \rightarrow \infty$, where according to 3.1 and 3.2 the electric field of the transverse modes vanishes, the two lower-frequency branches again describe purely mechanical phonons with vanishing phase velocities. The third branch describes photons propagating through the medium with phase velocity $c/\sqrt{\varepsilon_\infty}$.

The polariton dispersion branches of cubic crystals with an arbitrary number of atoms in the unit cell are derived from (4-23) by taking into account a corresponding number of polar modes in the dielectric function, (4-14).

Since the first successful light-scattering experiments on bulk polaritons in GaP /7/, the Raman effect has remained the principal experimental method giving information on these quasiparticles. We therefore discuss the corresponding techniques in more detail.

Energy and quasimomentum conservation are given by

$$\hbar\omega_i = \hbar\omega_s \pm \hbar\omega \quad , \qquad (4\text{-}26)$$

and

$$\hbar\underline{k}_i = \hbar\underline{k}_s \pm \hbar\underline{k} \quad , \qquad (4\text{-}27)$$

respectively, the positive signs describing Stokes and the negative signs Anti-Stokes processes; i stands for 'incident' and s for 'scattered', see 2.1. We recall (see 2.2) that in a cubic (optically isotropic) crystal the magnitude of the wave-vectors of the incident and scattered photons are given by $k_i = 2\pi n(\lambda_i)/\lambda_i$ and $k_s = 2\pi n(\lambda_s)/\lambda_s$, respectively; n is the refractive index of the medium, and λ_i and λ_s are the wavelengths in vacuum. For a right angle Raman scattering process with $\phi = \measuredangle(\underline{k}_i, \underline{k}_s) = \pi/2$ and an exciting frequency in the visible, we have seen that the magnitudes of the wave-vectors of the observed elementary excitations are of the order of $k \approx 10^5\ \mathrm{cm}^{-1}$, which corresponds to the phonon limit in polariton theory, see 2.2 and 3.3. Polaritons may therefore be observed by right angle scattering only when using exciting frequencies in the infrared (e.g. a CO_2 laser), where $\underline{k}_i$ is approximately one order of magnitude smaller. In the visible region when, for instance, an argon laser (488 nm, 414.5 nm), a Krypton laser (568.2 nm, 647.1 nm), or a He-Ne laser (632.8 nm) is used as the exciting light source, polaritons can be observed only by near-forward scattering. The high intensities and collimation of laser beams make it possible to attain well-defined scattering processes as sketched in Fig.5. The reduction of the scattering angle ϕ to near-forward scattering generally allows the observation of elementary excitations with wave-vectors of approximately one order of magnitude less. Furthermore, the variation of ϕ allows a continuous variation of k. The wave-vector relation for a Stokes process $\underline{k}_i = \underline{k}_s + \underline{k}$ derived from (4-27) leads to the following expression for the magnitude of $\underline{k}$:

$$k = (k_i^2 + k_s^2 - 2k_i k_s \cos\phi)^{1/2} \quad . \tag{4-28}$$

This can be written alternatively

$$k = \left[(k_i - k_s)^2 + 2k_i k_s (1-\cos\phi)\right]^{1/2} \quad . \tag{4-29}$$

Because the refractive index is $n = c(\partial k/\partial\omega)_{\omega_i}$ and $(\partial k/\partial\omega)_{\omega_i} \approx (k_i-k_s)/(\omega_i-\omega_s)$, it follows that

$$k = \left[(n^2/c^2)\omega^2 + 2k_i k_s (1-\cos\phi)\right]^{1/2} \tag{4-30}$$

taking energy conservation, (4-26), into account. (4-30) holds on-

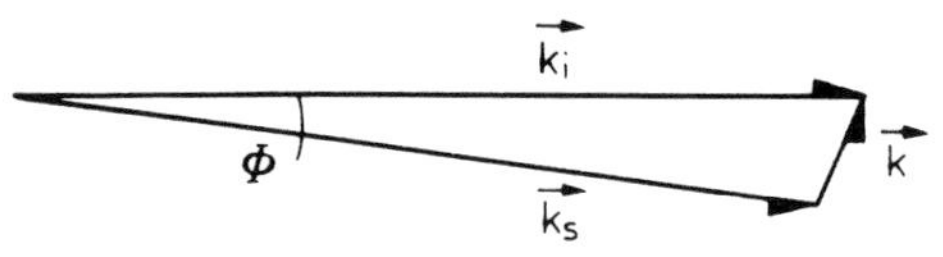

Fig.5 Near-forward scattering process, allowing the observation of polaritons when exciting laser frequencies in the visible region are used.

ly for $n(\omega_i) \approx n(\omega_s)$. A linear approximation where $n(\omega_s) \approx n(\omega_i) + (\partial k/\partial\omega)_{\omega_i}|\omega_i - \omega_s|$ has been discussed in /300/. If the refractive index cannot be regarded as constant within the region of the Raman spectrum, it is not possible to give a general prediction of the experimental situation on the basis of (4-30), (4-31) and (4-56) through (4-60); see also the discussion concerning (4-62). If we solve (4-30) together with the dispersion relation $\omega^2 = \omega^2(k)$, (3-31) or (4-23), the frequencies of the observed polaritons can be determined directly as a function of ϕ: $\omega^2 = \omega^2(\phi)$. Graphically, (4-30) determines hyperbolas for different ϕ in the ω,k diagram. The smallest wave-vectors are obtained for direct forward scattering where $\phi = 0$. In this case the two asymptotes of the hyperbolas are obtained. Their equations are

$$\omega = \pm(c/n)k \quad . \tag{4-31}$$

The intersection points of these 'limiting lines' with the polariton dispersion curves in the ω,k diagram determine the greatest frequency shifts which can be observed experimentally, see Fig.6. Then, for $\phi > 0$, the intersections of the hyperbolas with the dispersion curves determine the observed polariton frequencies. The limiting line with negative slope gives the same information but for negative $\underline{k}$ directions, see Fig.2 left-hand side.

The basic condition for observing bulk polaritons by light scattering is that there be simultaneous IR and Raman activity of the corresponding modes. Among the five cubic crystal classes the direction groups T (= 23) and T_d (= $\bar{4}3$ m) have no inversion center. Their triply degenerate modes (F modes) are consequently both IR- and Raman-active, so that polaritons may be directly observed. In 1965 Henry and Hopfield /7/ published the first successful experimental observation of phonon-polariton dispersion in GaP. The mate-

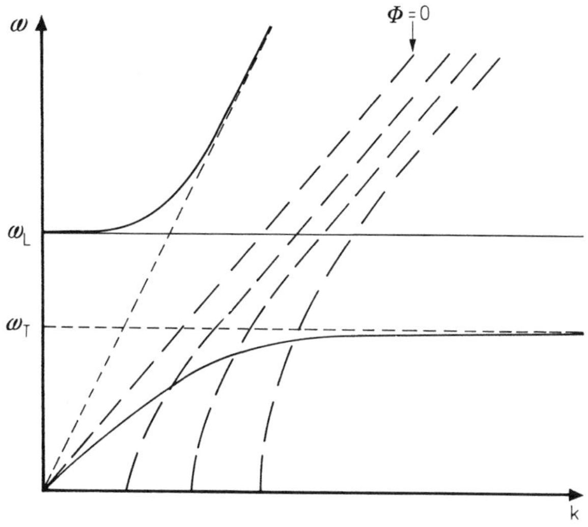

Fig.6 On the observation of polaritons in optically isotropic crystals by light scattering, see text.

rial belongs to the space group T_d (= F $\bar{4}$3 m). The exciting light source was a 35 mW He-Ne laser and the recording was performed photographically. The calculated dispersion curves and the experimental data points are reproduced in Fig.7. For most insulators $c/n(\omega_i) > c/\sqrt{\varepsilon_o}$ *), so that the limiting line, (4-31), has its intersection point with the lowest-frequency dispersion branch in the origin (Fig.6). In GaP, however, $c/n(\omega_i) < c/\sqrt{\varepsilon_o}$ *) because of a small band gap. This causes an intersection with the polariton branch at a finite frequency, $\omega \neq 0$.

As can be seen from (4-31), the maximum frequency shift observable depends on the refractive index of the material. Leite, Damen and Scott /82/ first showed this in ZnSe. Like GaP, ZnSe has only one polar mode. Furthermore, for this material too, the laser lines at 488, 514.5, 568, and 632.8 nm do not lie too far away from electron resonances, so the refractive index depends very much on the laser wavelength used for excitation. For this reason, the polariton shift observed for He-Ne laser excitation at 632.8 nm is more than 100 cm^{-1}, whereas an argon laser at 488 nm allows the observation

*) Note again that the limiting 'lines' exist only for the approximation $n(\omega_i) \approx n(\omega_s)$.

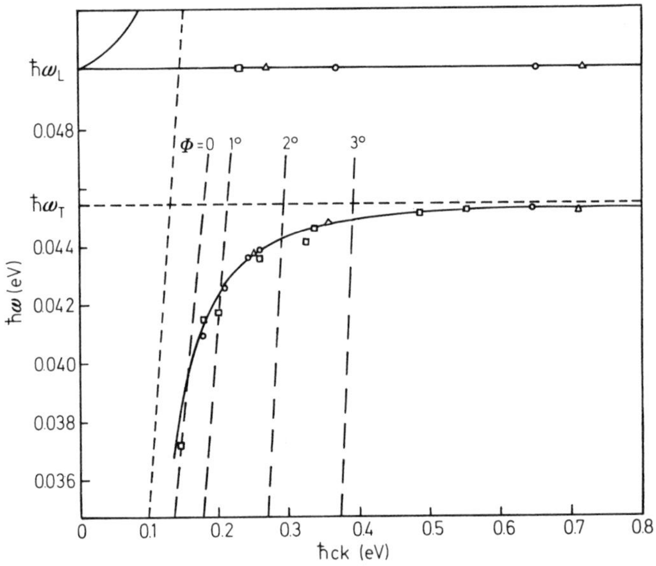

Fig.7 Calculated polariton dispersion curve and experimental data points for GaP, from /7/ (1 eV ≙ 8065 cm^{-1}).

of a shift of hardly more than 20 cm^{-1}, see also /71/.

The first cubic crystal with an inversion center in which polaritons were observed by Raman scattering was $KTaO_3$ /83/. The experiments were carried out at 8 K, where the material belongs to the space group O_h^1 (= Pm 3 m). A pulsed electric field of 250 to 2000 V was applied along a 2.4-mm crystal, periodically lifting the inversion center, so (periodically) inducing a Raman activity of the IR-active modes. The use of a chopper allowed the observation of polaritons associated with the TO phonon at 556 cm^{-1}. A frequency shift of 16 cm^{-1} down to 540 cm^{-1} for an internal scattering angle of $\phi \approx 2^o$ was recorded. The experimental technique has been described in detail by Fleury and Worlock /84/.

Further results concerning electric field-included Raman activity have been reported for $SrTiO_3$, /83/. This material, however, is no longer cubic. It belongs to the space group D_{4h}^{18} (= I 4/mcm) in the low-temperature phase.

Additional Literature

D'Andrea, A., Fornari, B., Mattei, G., Pagannone, M., Scrocco, M.: Light scattering from polaritons in $NaBrO_3$ and $NaClO_3$ /234/.

4.4 POLARITONS IN UNIAXIAL CRYSTALS

Uniaxial crystals are characterized by dielectric functions which are identical for two principal directions. These directions describe the 'optically isotropic' plane

$$\varepsilon_1(\omega) = \varepsilon_2(\omega) \equiv \varepsilon_\perp(\omega) \quad . \tag{4-32}$$

For the third principal direction perpendicular to this plane and referred to as the 'optic axis', the dielectric function is different:

$$\varepsilon_3(\omega) \equiv \varepsilon_\parallel(\omega) \neq \varepsilon_\perp(\omega) \quad . \tag{4-33}$$

The symbols $\perp$ and $\parallel$ denote directions perpendicular and parallel to the optic axis, respectively ($\alpha = 3$). The number of polar modes for the two principal orthogonal directions in the optically isotropic plane are identical: $n_1 = n_2 = n_\perp$. For an arbitrary direction $\underline{s}$ of the wave-vector $\underline{k}$, the three vectors $\underline{Q}$, $\underline{E}$ and $\underline{P}$ can be split into linearly independent 'ordinary' and 'extraordinary' components lying perpendicular to or in the plane determined by $\underline{s}$ and the optic axis, respectively:

$$\underline{Q} = \begin{pmatrix} \underline{Q}_o \\ \underline{Q}_{eo} \end{pmatrix} = \begin{pmatrix} Q_{\perp 1} \cdots Q_{\perp n_1} \\ Q_{\perp n_1+1} \cdots Q_{\perp n_2}, Q_{\parallel n_2+1} \cdots Q_{\parallel n} \end{pmatrix} \quad .$$

Note that $n_1 = n_2 \neq n_3$ and $n_1 + n_2 + n_3 = n$. We furthermore define

$$\underline{E} = \begin{pmatrix} \underline{E}_o \\ \underline{E}_{eo} \end{pmatrix} = \begin{pmatrix} E_\perp \\ E_\perp \, , \, E_\parallel \end{pmatrix}$$

and

$$\underline{P} = \begin{pmatrix} \underline{P}_o \\ \underline{P}_{eo} \end{pmatrix} = \begin{pmatrix} P_\perp \\ P_\perp \, , \, P_\parallel \end{pmatrix} \quad .$$

If we split the coefficient matrices $B^{\mu\nu}$ in (4-7) and (4-8) or (4-10) and (4-11) into two parts, $B_o^{\mu\nu}$ and $B_{eo}^{\mu\nu}$, with respect to the ordinary and extraordinary components of the vectors $\underline{Q}$, $\underline{E}$ and $\underline{P}$, we obtain two different fundamental sets of equations describing 'ordinary' and 'extraordinary' polaritons. The two corresponding relations for the electric-field components are

$$\underline{E}_o = 4\pi(n^2-1)^{-1}\,\underline{P}_o \quad , \tag{4-34}$$

and

$$\underline{E}_{eo} = 4\pi(n^2-1)^{-1}[\underline{P}_{eo}-n^2\underline{s}(\underline{s}\cdot\underline{P}_{eo})] \quad . \tag{4-35}$$

This can easily be derived from (4-9). Since P_o is always perpendicular to the plane determined by $\underline{s}$ and the optic axis, the second term in the bracket of (4-9) vanishes for ordinary polaritons. Ordinary polaritons thus do not show directional dispersion because $\underline{s}$ no longer appears in (4-34).

The set of equations describing extraordinary polaritons, on the other hand, includes (4-35) for the electric field. In this relation $(\underline{s}\cdot P_{eo})$ will generally not vanish since the only assumption made was that $\underline{P}_{eo} = (P_\perp, P_\parallel)$ lies in the plane determined by $\underline{s}$ and the optic axis. Accordingly, extraordinary polaritons become of mixed type, being partly longitudinal and partly transverse for arbitrary wave-vector directions.

In terms of group theory the modes with displacement vector components lying only perpendicular to the optic axis are said to be twofold degenerate. Ordinary polaritons with $\underline{Q}_o = (Q_{\perp 1}\ldots Q_{\perp n_1})$ are thus associated for all wave-vector directions with exactly transverse phonons of E type. The extraordinary displacement vector on the other hand is $\underline{Q}_{eo} = (Q_{\perp n_1+1}\ldots Q_{\perp n_2}, Q_{\parallel n_2+1}\ldots Q_{\parallel n})$. This implies quasinormal coordinate components both perpendicular and parallel to the optic axis. Extraordinary polaritons for arbitrary directions $\underline{s}$ therefore are associated with phonons of mixed type E + A, the A-type ones referring to totally symmetric components with the index $\parallel$. We shall later see that only for wave-vectors propagating in the optically isotropic plane do extraordinary polaritons show pure A character with lattice displacements parallel to the optic axis (z). On the other hand, for $\underline{s}\parallel z$ extraordinary polaritons degenerate with the ordinary ones. Both are then of pure E(TO) type. For more detailed discussions, see 4.7.

If the fundamental equations for the ordinary and extraordinary components of $\underline{Q}$, $\underline{E}$ and $\underline{P}$ are solved separately by the procedure described in 4.2, different dispersion relations are obtained for the ordinary and extraordinary polaritons. The result, however, can also be seen directly from the general relation (4-21), taking into account (4-32) and (4-33). Eq.(4-21) splits into two factors:

$$(\omega^2\varepsilon_\perp(\omega)-c^2k^2)\left[\varepsilon_\perp(\omega)(\omega^2\varepsilon_\parallel(\omega)-c^2k^2)s_\perp^2+\varepsilon_\parallel(\omega)(\omega^2\varepsilon_\perp(\omega)-c^2k^2)s_\parallel^2\right] = 0 \quad . \tag{4-36}$$

The zeros of the first factor, depending on the magnitude of the wave-vector but not its direction, describe the <u>ordinary polaritons.</u> The second bracket on the other hand contains $\varepsilon_\perp(\omega)$ as well as $\varepsilon_\parallel(\omega)$, the magnitude of the wave-vector $\underline{k}$, and its direction $\underline{s} = (s_\perp, s_\parallel)$ where $s_\perp = (s_1^2 + s_2^2)^{1/2}$ and $s_\parallel \equiv s_3$. This bracket correspondingly gives the dispersion relation of <u>extraordinary polaritons.</u> When the angle θ between the optic axis and $\underline{s}$ is introduced, $s_\perp^2 = \sin^2\theta$ and $s_\parallel^2 = \cos^2\theta$. The equation can then be written in the familiar form:

$$\varepsilon(\omega,\theta) = c^2k^2/\omega^2 = \frac{\varepsilon_\perp(\omega)\varepsilon_\parallel(\omega)}{\varepsilon_\perp(\omega)\sin^2\theta + \varepsilon_\parallel(\omega)\cos^2\theta} \quad . \tag{4-37}$$

The dielectric functions $\varepsilon_\parallel(\omega)$ and $\varepsilon_\perp(\omega)$ are explicitly determined by two Kurosawa relations containing the LO and TO frequencies of the polar modes with lattice displacements parallel and perpendicular to the optic axis, respectively. In terms of group theory this means all LO and TO frequencies of polar A and E phonons.

Among the uniaxial crystal classes in the two direction groups D_{2d} (= $\bar{4}$2 m) and S_4 (= $\bar{4}$) there are no totally symmetric polar phonons. In these cases the modes of symmetries B_2 and B, respectively, have dipole moments parallel to the optic axis. In order to avoid confusion, which could easily arise if these exceptions are always pointed out in all derivations throughout, we are henceforth going to discuss polar 'A' and 'E' phonons in uniaxial crystals only. The reader may transpose the notations for the crystal classes D_{2d} (= $\bar{4}2$ m) and S_4 (= $\bar{4}$) if necessary.

4.5 EXPERIMENTAL ARRANGEMENTS

The principle of the experimental technique for the observation of polaritons by light scattering has already been described in 4.3. The scattered light is recorded in near-forward directions at small angles ϕ away from the direction of the incident laser beam. Variation of ϕ is used in order to vary the magnitude of the wave-vector of the observed polaritons. From the dispersion relation of extraordinary polaritons, (4-37), it can be seen that the frequency ω is a function of k as well as of its direction $\underline{s}$. Experimental investigations in uniaxial crystals therefore require more careful experimental techniques, allowing separate observation of directional dispersion and of dispersion due to the magnitude of $\underline{k}$.

In many papers on Raman scattering by polaritons the experimental arrangements have been described rather indicatively so that it is hardly possible to make a quantitative comparison with re-recorded data. We therefore briefly discuss some basic experimental problems and present a simple and useful scattering arrangement by Claus /85/.

When the exciting laser beam is weakly focused into the sample, the radiation scattered by polaritons with certain wave-vector magnitudes is emitted in forward directions conically around the incident laser beam. If ϕ, as before, denotes the angle between $\underline{k}_i$ and $\underline{k}_s$ in the wave-vector diagram $\underline{k}_i = \underline{k}_s + \underline{k}$ (see (4-27)) 2ϕ is the vertex angle of such a cone (inside the sample). The experimentally observed scattered radiation leaves the sample in directions between two cones of vertex angles $2(\phi'-(1/2)\Delta\phi)$ and $2(\phi'+(1/2)\Delta\phi)$. The scattering angle inside the medium (ϕ) has to be calculated from the corresponding angle ϕ' observed outside. For refractive indices $n > 1$ ϕ is always smaller than ϕ', which is advantageous for the experiment. Reduction of $\Delta\phi$ will allow a more precise determination of the average wave-vector of the observed polaritons, while a greater divergence angle $\Delta\phi$ gives a more intense spectrum. The most suitable value of $\Delta\phi$ must be chosen in each case with respect to these two effects, the material, size of sample, laser output, electronic recording system, etc. Especially in regions of k space with strong dispersion, the '$\Delta\phi$ error' causes a remarkable broadening of the observed Raman lines /86/ because polaritons corresponding to different wave-vectors are recorded simultaneously. A scattering arrangement showing the '$\Delta\phi$ error' is sketched in Fig.8a. Furthermore, the extension of the crystal sample along the exciting laser beam causes an error in ϕ, too, as illustrated in Fig.8b. In this case only the use of a long-focus projection lens (1) or a thin sample will allow more accurate determination of the polariton wave-vectors. The technique described above is the most efficient and can be used for cubic crystals where anisotropy effects do not have to be taken into account.

In uniaxial and biaxial crystals the phonon frequencies in general depend on the direction of the wave-vector because of the anisotropy of the materials. Since for different scattering angles ϕ the polariton wave-vector also changes its direction drastically, only the use of a 'scattering plane' which is at the same time an optically isotropic plane of the crystal will allow observation of pure polariton dispersion (due to the magnitude of the wave-vector). 'Scattering plane' means that all wave-vector triangles are arranged parallel and not conically around the incident laser beam. In Fig.9a series of typical scattering triangles: $\underline{k}_i = \underline{k}_s + \underline{k}$ have been drawn to scale in order to show the effect. It can easily be seen that variation of ϕ from 0.7° to 4.0° changes both the magnitude and the direction of $\underline{k}$: the angle θ changes from 68° to 27°. Thus, when optically anisotropic scattering planes are used, the shifts of Raman lines observed will be caused by both polariton and directional dispersion. Extraordinary polaritons, see 4.4 were first observed in 1970 in uniaxial $LiIO_3$ by Claus /85, 86/. Fig.10 shows a corresponding experiment for $LiNbO_3$. Totally symmetric polaritons associated with the A_1(TO) phonon at 633 cm^{-1} were recorded in the optically isotropic plane, Fig.10a (no directional dispersion). For increasing scattering angles ϕ from 0.7°, the polariton moves toward higher wave numbers and finally coincides with the position of the pure A_1(TO) phonon at 633 cm^{-1}. Note that the scattering cross-section of the polariton increases simultaneously. Fig.10b shows the same experiment done with a scattering plane containing the optic axis. In this case we are observing directionally depen-

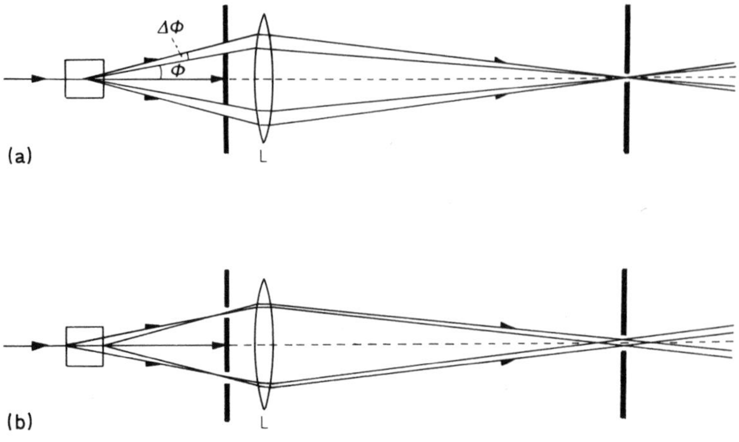

Fig.8 a) Error due to the finite divergence of the observed beams of scattered radiation; b) error due to the extension of the sample along the optical axis of the system, from /85/.

dent extraordinary polaritons. For increasing scattering angles starting with $\phi = 1.0^{\circ}$, the polariton again increases in frequency but moves only toward the position of an E(TO) phonon at 582 cm^{-1}. The scattering intensity decreases because the polariton becomes increasingly E-type with increasing ϕ. We shall later see (Figs.16 and 17) that the two phonons, A_1(TO) at 633 cm^{-1} and E(TO) at 582 cm^{-1}, are coupled by a directional dispersion branch. The scans in Fig.10a and b correspond exactly to the scattering triangles sketched in Fig.9. The dashed vertical line in Fig.9 indicates the direction of the optic axis for the experiments in Fig.10b.

An arrangement that readily allows the observation of radiation from parallel scattering planes is illustrated in Figs.11 and 12. Fig.11a is a vertical and 11b a horizontal projection. The entrance slit of the double monochromator I is assumed to be vertical. The exciting laser beam that passes through the optical parts of the system, labeled in alphabetic sequence, starts as a parallel bundle of rays entering the system on the left of Fig.11. A is a glass plate which serves to collect some light for the reference signal when a two-channel recording system is used. B is a cylindrical lens with its cylinder axis lying horizontally. The beam is diverged to a vertical band by this lens which may in principle be either positive or negative. The total distance to the crystal sample E

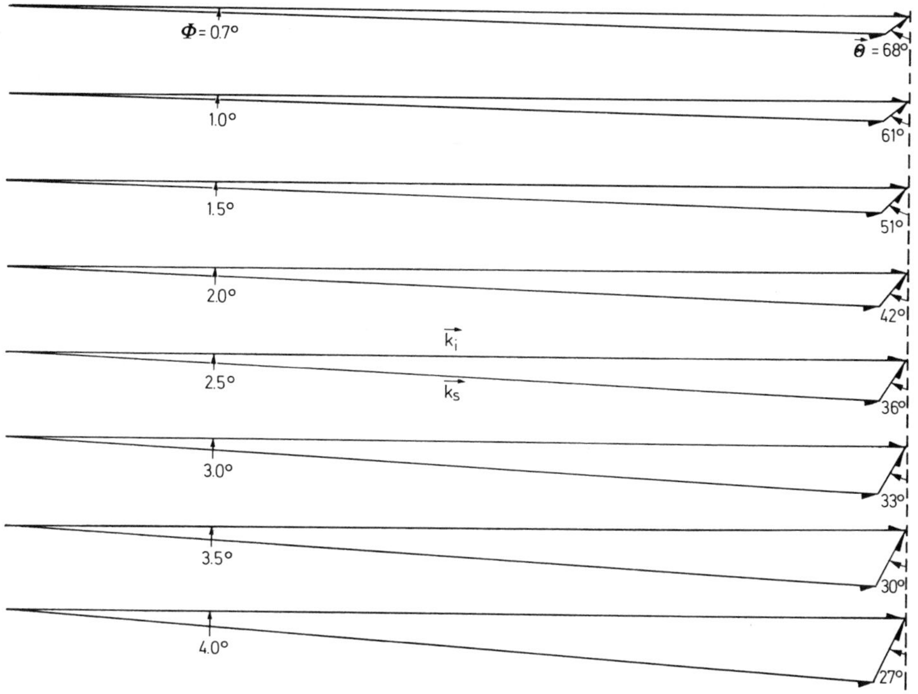

Fig.9 Scattering triangles $\underline{k}_i = \underline{k}_s + \underline{k}$ corresponding to polaritons associated with the A_1(TO)-phonon at 633 cm^{-1} in $LiNbO_3$ for certain forward-scattering geometries. The diagrams are plotted to scale, see /75/.

should be chosen so that the height of the exciting beam is not larger than the sample at position E and the length of the entrance slit of the double monochromator at I. The shape of the laser beam in the vertical plane of the system is sketched in Fig.11b up to part F, which is a vertical slit limiting the divergence $\Delta\phi$ of the observed scattered radiation. C is a thick glass plate allowing a parallel displacement of the laser beam in the horizontal plane. The magnitude of the displacement can be changed by simply rotating the glass plate around a vertical axis. D is another cylindrical lens with the axis of the cylinder perpendicular to the plane of Fig.11a. The laser is focused into the vertical focal line of this lens in such a way that the angle ϕ of forward scattering is simultaneously changed when the glass block C is rotated. By neglecting the spherical aberration of the lens, the position of the focal line is made independent of ϕ. For scattering angles $\phi > 0$ the ex-

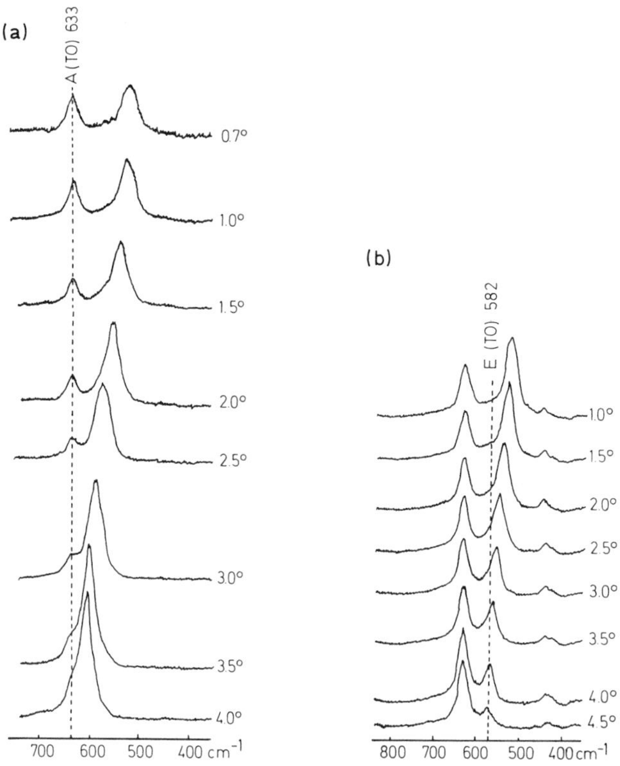

Fig.10 a) Polaritons associated with the A_1(TO) phonon at 633 cm^{-1} in $LiNbO_3$. Scattering triangles are lying in the optically isotropic plane. b) Extraordinary polaritons associated with the corresponding extraordinary phonons when using an optically anisotropic scattering plane. This spectra series corresponds to the wave-vector triangles sketched in Fig.9, from /284/.

citing laser light cannot directly pass F, so any polarizer can be used before the entrance slit I without being destroyed. Owing to the divergence of the incident laser radiation, direct forward scattering ($\phi = 0$) due to the diagonal elements of the Raman tensor is in principle also possible. The divergence avoids local destruction of the gratings and mirrors in the monochromator, see for instance /85/ and Fig.32. For straight forward scattering, however, crystal polarizers necessarily have to be used before the entrance slit I. The lens H projects the scattering volume on this slit. Because of the design of the whole arrangement, the image appears as a bright bar which can easily be made to coincide with the slit.

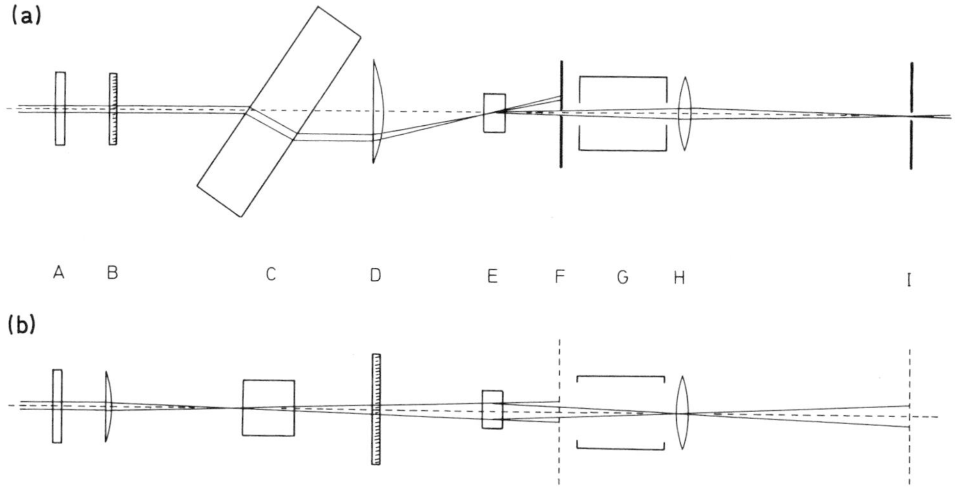

Fig.11 Arrangement for observing scattered light from polaritons in anisotropic crystals. a) vertical projection; b) horizontal projection, from /85/.

However, scattered light from polaritons does not leave the specimen only at certain angles ϕ in the horizontal plane of Fig.11a. Because of the finite height of the slits F and I, light at different scattering angles in the vertical plane (Fig.11b) will also be recorded at the same time. In order to avoid a broadening of the Raman lines due to this effect, which in uniaxial crystals can involve shifts due to the anisotropy as well as polariton shifts, a Venetian blind (sketched in Fig.12) is introduced at position G. The horizontal slats of the blind restrict the divergence of the observed scattered radiation in the vertical plane to angles less than θ away from the optic axis of the system.

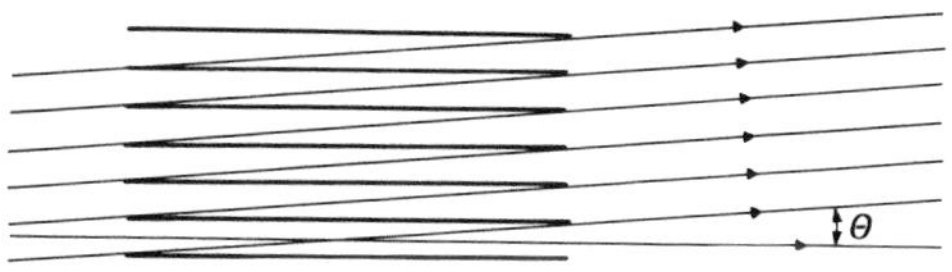

Fig.12 Venetian blind restricting the divergence of the observed radiation in the vertical plane to angles less than Θ, from /85/.

Additional Literature

Ahrens, K.H.F., Schaack, G., Unger, B.: Instrumental Effects in the Raman-lineshapes of Polaritons /352/.

4.6 DIRECTIONAL DISPERSION OF EXTRAORDINARY PHONONS IN UNIAXIAL CRYSTALS (OBLIQUE PHONONS)

The dispersion relation of extraordinary polaritons in uniaxial crystals (4-37), which was derived in 4.4, can be written

$$\omega^2/c^2k^2 = (\varepsilon_{\perp}(\omega)\sin^2\theta + \varepsilon_{\parallel}(\omega)\cos^2\theta)/\varepsilon_{\perp}(\omega)\varepsilon_{\parallel}(\omega) \quad . \tag{4-38}$$

It can be seen directly from (4-38) that the equation for $k \to \infty$ reduces to

$$tg^2\theta = -\varepsilon_{\parallel}(\omega)/\varepsilon_{\perp}(\omega) \quad . \tag{4-39}$$

There is no dispersion due to the magnitude of the wave-vector but there is due to its direction. Eq.(4-39) thus describes the directional dependence of extraordinary phonons in the region $10^5 \lesssim k \lesssim 10^7\ \mathrm{cm}^{-1}$. On taking the Kurosawa relation (4-14) into account, (4-39) becomes

$$\sin^2\theta\cdot\varepsilon_{\perp\infty}\prod_{i=1}^{n_\perp}(\omega^2_{\perp Li}-\omega^2)\prod_{j=1}^{n_\parallel}(\omega^2_{\parallel Tj}-\omega^2) +$$

$$+\cos^2\theta\cdot\varepsilon_{\parallel\infty}\prod_{j=1}^{n_\parallel}(\omega^2_{\parallel Lj}-\omega^2)\prod_{i=1}^{n_\perp}(\omega^2_{\perp Ti}-\omega^2) = 0 \quad , \tag{4-40}$$

where $n_\perp$ and $n_\parallel$ stand for the number of polar modes for wave-vector directions in the optical plane and parallel to the optic axis, respectively. The total number of modes observed is $n_\perp + n_\parallel = n$. The dependence of the phonon frequencies of extraordinary modes on the angle θ in (4-39) implies that a mode showing a certain frequency when the wave-vector propagates parallel to the optic axis (z) continuously changes frequency as the wave-vector is changed, until it finally propagates perpendicular to z. The frequency observed for $k \perp z$ is that of another fundamental mode. The latter phonon is thus said to be coupled to the first one by 'directional dispersion'.

The coupling of extraordinary phonon modes in uniaxial crystals between $\theta = 0$ and $\theta = \pi/2$ can be of four fundamental types: (1) LO $\leftrightarrow$ TO; (2) LO $\leftrightarrow$ LO; (3) TO $\leftrightarrow$ LO, and (4) TO $\leftrightarrow$ TO. For $\theta = 0$, the modes can be of symmetry type A(LO) or E(TO), and for $\theta = \pi/2$

of type A(TO) or E(LO). The E(TO) modes for $\theta = \pi/2$ are ordinary phonons. Table 6 shows the possible couplings in a schematic form. The fact that the coupling is of the form LO ↔ LO, TO ↔ TO, A ↔ A, or E ↔ E does not imply that LO, TO, A, or E character is also presented for intermediate θ values $0 < \theta < \pi/2$ /72/. For example, longitudinal phonons, which are only Raman-active for $\theta = 0$ and $\theta = \pi/2$, may also be infrared-active in the intermediate region because they have some TO character /87/. This leads us to the problem of the directional dependence of infrared oscillator parameters for extraordinary phonons.

	$\underline{k} \parallel z$		$\underline{k} \perp z$	
	$\theta = 0$		$\theta = \frac{\pi}{2}$	
	Frequency-symbol	Symmetry	Symmetry	Frequency-symbol
Extraordinary phonons	$\omega_{L\parallel}$ $\omega_{T\perp}$	A(LO) E(TO)	A(TO) E(LO)	$\omega_{T\parallel}$ $\omega_{L\perp}$
Ordinary phonons	$\omega_{T\perp}$	E(TO)	E(TO)	$\omega_{T\perp}$

Tab.6 Possibilities of phonon coupling due to directional dispersion in uniaxial crystals. z denotes the optic axis.

According to (4-18), the mode strength of the k-th transverse mode for $\theta = 0$ is given by

$$S_k = \frac{\varepsilon_{\perp\infty} \prod_{j=1}^{n_\perp} (\omega_{\perp Lj}^2 - \omega_{\perp Tk}^2)}{\omega_{\perp Tk}^2 \prod_{\substack{j=1 \\ j\neq k}}^{n_\perp} (\omega_{\perp Tj}^2 - \omega_{\perp Tk}^2)} \quad . \tag{4-41}$$

The relation can be written correspondingly for an extraordinary TO mode propagating perpendicular to the optic axis. The index $\perp$ has only to be replaced by $\parallel$. For the lattice displacements $\underline{Q}_i$, holds in this case $\underline{Q}_i \parallel z$. These two relations describe the mode strengths for two fixed principal directions. The oscillator strength for an extraordinary phonon with an arbitrary wave-vector direction can be derived as follows /88/.

The frequencies $\omega = \omega_i(\theta)$ of the $n_\parallel + n_\perp = n$ polar extraordinary-phonon modes for a certain angle θ between the optic axis and the wave-vector are determined by (4-40), i.e. the zeros of the denominator of the directionally dependent dielectric function (4-37). This denominator can be written alternatively:

$$(\varepsilon_{\perp\infty} \sin^2\theta + \varepsilon_{\parallel\infty} \cos^2\theta) \prod_{i=1}^{n} (\omega_i^2(\theta) - \omega^2) \quad .$$

If we use this notation, the dielectric function becomes

$$\varepsilon(\omega,\theta) = \frac{\varepsilon_{\perp\infty}\varepsilon_{\parallel\infty} \prod_{j=1}^{n_\perp} (\omega_{\perp Lj}^2 - \omega^2) \prod_{\ell=1}^{n_\parallel} (\omega_{\parallel L\ell}^2 - \omega^2)}{(\varepsilon_{\perp\infty} \sin^2\theta + \varepsilon_{\parallel\infty} \cos^2\theta) \prod_{i=1}^{n} (\omega_i^2(\theta) - \omega^2)} \quad . \tag{4-42}$$

$\varepsilon_\perp(\omega)$ and $\varepsilon_\parallel(\omega)$ have been replaced by the corresponding Kurosawa relations (4-14). Alternatively, the dielectric function in question can be written, see (4-13)

$$\varepsilon(\omega,\theta) = \varepsilon_\infty(\theta) + \sum_{i=1}^{n} \frac{S_i \omega_i^2(\theta)}{\omega_i^2(\theta) - \omega^2} \quad . \tag{4-43}$$

From (4-42) and (4-43) the directional dependence of the mode strength can easily be derived for $\omega \to \omega_k(\theta)$

$$S_k(\theta) = \frac{\varepsilon_{\perp\infty}\varepsilon_{\parallel\infty} \prod_{j=1}^{n_\perp} (\omega_k^2(\theta) - \omega_{\perp Lj}^2) \prod_{\ell=1}^{n_\parallel} (\omega_k^2(\theta) - \omega_{\parallel L\ell}^2)}{(\varepsilon_{\perp\infty} \sin^2\theta + \varepsilon_{\parallel\infty} \cos^2\theta)\,\omega_k^2(\theta) \prod_{\substack{i=1 \\ i \neq k}}^{n} (\omega_k^2(\theta) - \omega_i^2(\theta))} \quad . \tag{4-44}$$

In the limiting case $\theta = 0$, where the wave-vector propagates parallel to the optic axis, the modes are either exactly transverse: $\omega_T(0^\circ) = \omega_{\perp Tj}$, $(j = 1 \ldots n_\perp)$ (E(TO) modes), or exactly longitudinal: $\omega_{Lk}(0^\circ) = \omega_{\parallel L\ell}$, $(\ell = 1 \ldots n_\parallel)$ (A(LO) modes). In the first case (4-44) reduces to

$$S_{kT}(0^\circ) = \frac{\varepsilon_{\perp\infty} \prod_{j=1}^{n_\perp} (\omega_{\perp Tk}^2 - \omega_{\perp Lj}^2)}{\omega_{\perp Tk}^2 \prod_{\substack{j=1 \\ j \neq k}}^{n_\perp} (\omega_{\perp Tk}^2 - \omega_{\perp Tj}^2)} \quad . \tag{4-45}$$

All factors with indices $\parallel$ in the numerator and denominator are cancelled. Eq. (4-45) is identical with (4-41). In the second case, on the other hand, one of the factors in the numerator always becomes zero. This implies that the mode strengths of all longitudinal waves vanish.

An analogous result is obtained for $\theta = \pi/2$. The purely transverse extraordinary modes are then of totally symmetric type: $\omega_{Tk}(\pi/2) = \omega_{\parallel T\ell}$, $(\ell = 1 \ldots n_\parallel)$ and the exactly longitudinal vibrations are E(LO) modes: $\omega_{Lk}(\pi/2) = \omega_{\perp Lj}$, $(j = 1 \ldots n_\perp)$. The equation corresponding to (4-45) is obtained in this case by simply replacing $\perp$ by $\parallel$.

Finally, we derive the directional dependence of the high-frequency dielectric constant from (4-42) and (4-43) for $\omega \to \infty$ as

$$\varepsilon_\infty(\theta) = \frac{\varepsilon_{\perp\infty}\, \varepsilon_{\parallel\infty}}{\varepsilon_{\perp\infty} \sin^2\theta + \varepsilon_{\parallel\infty} \cos^2\theta} \quad . \tag{4-46}$$

Eqs.(4-44) and (4-46) describe the directional dependence of the oscillator parameters when damping is neglected. A quantitative comparison with experimental data for α quartz is given in /88/.

If we wish to determine phonon frequencies for a certain wave-vector direction, (4-40) has to be solved. For crystals with a large number of atoms in the unit cell this can be done numerically only by computer. We shall therefore discuss some approximations of (4-40) that allow the analytic calculation of certain isolated directional dispersion branches in polyatomic crystals.

The simplest expression hitherto used in the literature is the Poulet-Loudon approximation /31, 89/:

$$\omega^2(\theta) = \omega_\alpha^2 \cos^2\theta + \omega_\beta^2 \sin^2\theta \quad , \tag{4-47}$$

where ω_α and ω_β are the phonon frequencies for wave-vector directions respectively parallel ($\theta = 0^\circ$) and perpendicular ($\theta = \pi/2$) to the optic axis, and are coupled by the directional dispersion branch in question. For $\theta = 0^\circ$, ω_α can be either an A(LO) or E(TO) phonon, and for $\theta = \pi/2$ ω_β can be either an A(TO) or an E(LO) phonon, see Table 6. The Poulet-Loudon approximation is obviously derived from the rigorous relation (4-40) for $\varepsilon_{\perp\infty} \approx \varepsilon_{\parallel\infty}$ when the directional dispersion of all the other branches is neglected, i.e. their frequencies for $\theta = 0$ and $\theta = \pi/2$ are assumed to be equal. The Poulet-Loudon approximation may be applied to extraordinary polar branches with small dispersion /90, 91/. If the directional dispersion is rather strong (dominating anisotropy) better approximations have to be used.

A linear approximation /92/ of (4-40) is obtained when in all factors except $(\omega_\alpha^2-\omega^2)$ and $(\omega_\beta^2-\omega^2)$ the frequencies ω are replaced by an average value

$$\bar{\omega} = (1/2)(\omega_\alpha+\omega_\beta) \quad . \tag{4-48}$$

$\bar{\omega}$ could equally well be chosen as $\bar{\omega} = \sqrt{\omega_\alpha\omega_\beta}$ or $\bar{\omega}^2 = (1/2)(\omega_\alpha^2+\omega_\beta^2)$. The type of approximation for the average value has hardly any influence on the result. According to Table 6, ω_α can be any one of the frequencies $\omega_{\perp Ti}$ (as for instance: $\omega_\alpha \equiv \omega_{\perp TI}$) or $\omega_{\parallel Lj}$ ($\omega_\alpha \equiv \omega_{\parallel LJ}$). We can similarly introduce $\omega_\beta \equiv \omega_{\parallel TJ}$ or $\omega_\beta \equiv \omega_{\perp LI}$. The four principal types of extraordinary phonon couplings which can take place are described by the four relations:

$$d_{\parallel}^{(\mu)} \cos^2\theta\,(\omega_\alpha^2-\omega^2)+d_{\perp}^{(\nu)} \sin^2\theta\,(\omega_\beta^2-\omega^2) = 0 \tag{4-49}$$

where $\mu,\nu = 1, 2$. The set of constants is

$$d_{\parallel}^{(1)} = \varepsilon_{\parallel\infty} \prod_{\substack{i=1\\ i\neq I}}^{n_\perp} (\omega_{\perp Ti}^2-\bar{\omega}^2) \prod_{j=1}^{n_\parallel} (\omega_{\parallel Lj}^2-\bar{\omega}^2) \quad ,$$

$$d_{\parallel}^{(2)} = \varepsilon_{\parallel\infty} \prod_{i=1}^{n_\perp} (\omega_{\perp Ti}^2-\bar{\omega}^2) \prod_{\substack{j=1\\ j\neq J}}^{n_\parallel} (\omega_{\parallel Lj}^2-\bar{\omega}^2) \quad ,$$

$$d_{\perp}^{(1)} = \varepsilon_{\perp\infty} \prod_{i=1}^{n_\perp} (\omega_{\perp Li}^2-\bar{\omega}^2) \prod_{\substack{j=1\\ j\neq J}}^{n_\parallel} (\omega_{\parallel Tj}^2-\bar{\omega}^2) \quad ,$$

$$d_{\perp}^{(2)} = \varepsilon_{\perp\infty} \prod_{\substack{i=1\\ i\neq I}}^{n_\perp} (\omega_{\perp Li}^2-\bar{\omega}^2) \prod_{j=1}^{n_\parallel} (\omega_{\parallel Tj}^2-\bar{\omega}^2) \quad .$$

By solving (4-49) we obtain the following four equations for $\omega^2 = \omega^2(\theta)$

$$\omega^2(\theta) = \frac{d_{\parallel}^{(\mu)}\omega_\alpha^2 \cos^2\theta+d_{\perp}^{(\nu)}\omega_\beta^2 \sin^2\theta}{d_{\parallel}^{(\mu)} \cos^2\theta+d_{\perp}^{(\nu)} \sin^2\theta} \quad . \tag{4-50}$$

In the linear approximation n-2 coupled oscillators are thus formally treated as decoupled. The coupling is taken into account by introducing the constant coefficients $d_{\parallel}^{(\mu)}$ and $d_{\perp}^{(\nu)}$. Eq.(4-50) can be reduced to the Poulet-Loudon approximation for $d_{\parallel}^{(\mu)} = d_{\perp}^{(\nu)}$.

Merten and Lamprecht /92/ have shown that the linear approximation represents great progress compared with the Poulet-Loudon approximation although the calculating effort is not much greater. Numerical results for two-directional dispersion branches of α quartz are shown in Fig.13.

When two branches with strong dispersion lie quite close together

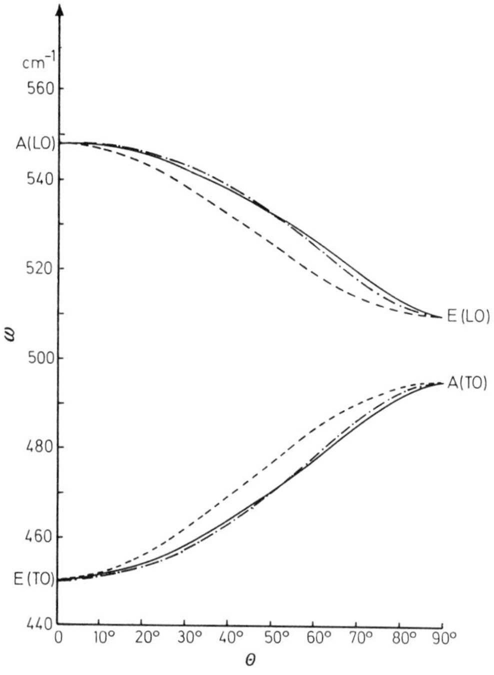

Fig.13 Directional dispersion of two extraordinary polar phonon branches in α quartz:
——————— : rigorous solution, Eq.(4-40)
-------- : Poulet-Loudon approximation, Eq.(4-47)
—·—·—·— : linear approximation, Eq.(4-50), from /92/.
θ denotes the angle between the wave-vector and the optic axis.

the accuracy of the calculation can be further improved by a quadratic approximation. In this approximation ω remains variable in two factors of (4-40) corresponding to the two branches in question. In analogy with (4-49), we then have

$$D_{\parallel}\cos^2\theta(\omega_{\alpha 1}^2-\omega^2)(\omega_{\alpha 2}^2-\omega^2)+D_{\perp}\sin^2\theta(\omega_{\beta 1}^2-\omega^2)(\omega_{\beta 2}^2-\omega^2) = 0 \qquad (4\text{-}51)$$

where the pairs of frequencies $(\omega_{\alpha 1}, \omega_{\beta 1})$ and $(\omega_{\alpha 2}, \omega_{\beta 2})$ determine the branches in the same way as in the linear approximation. The coefficients are

$$D_{\parallel} = \varepsilon_{\parallel\infty} \prod_{\substack{\sigma=1 \\ \sigma\neq\alpha_1 \\ \sigma\neq\alpha_2}}^{n} (\omega_{\sigma}^2-\bar{\omega}_{\parallel}^2)$$

and

$$D_{\perp} = \varepsilon_{\perp\infty} \prod_{\substack{\rho=1 \\ \rho\neq\beta_1 \\ \rho\neq\beta_2}}^{n} (\omega_\rho^2-\overline{\omega}_\perp^2) \quad .$$

Note that $n = n_\perp + n_\parallel$, so that the frequencies $\omega_{\perp Ti}$ and $\omega_{\parallel Lj}$ in (4-40) simply have to be arranged with respect to their magnitude and counted as ω_σ, while the $\omega_{\parallel Tj}$ and $\omega_{\perp Li}$ are similarly counted as ω_ρ. Eq.(4-51) is quadratic in ω^2 and can still be solved rigorously without difficulty.

Experimental investigations of the directional dispersion of extraordinary phonons by light scattering with exciting lines in the visible can be carried out by both rectangular and backward scattering. In both cases the magnitude of the wave-vectors of the observed excitations are large compared with those in the polariton region. In practice, however, the right-angle scattering technique has been shown to be somewhat disadvantageous. Reflections of the laser light on the second surface inside the specimen (in the direction of the exciting laser beam) generate Raman scattering of phonons at an angle $\theta' = (\pi/2-\theta)$ to the optic axis. This means that, for instance, the $\theta = 40^\circ$ spectrum will always be somewhat mixed up with the $\theta = 50^\circ$ spectrum. This effect can make the interpretation of the experimental data very difficult. The use of backward-scattering technique eliminates this disadvantage because the corresponding reflections then give rise to the $\theta' = (\pi-\theta)$ spectrum which is obviously identical with the θ spectrum.

An experimental setup is sketched in Fig.14. The incident laser beam comes from the top (arrow) and the back-scattered radiation is turned to the horizontal by a mirror. The wave-vector diagram for the observed scattering process is shown in the lower part of the figure. It is obvious that the wave-vector direction of the observed phonons can be made to propagate perpendicular to one of the surfaces when $\underline{k}_i$ and $\underline{k}_s$ are propagating symmetrically at the same angle to this surface. Rigorously, however, the magnitudes of the wave-vectors and the different refractive indices have to be taken into account. Furthermore, it must be noted that the direction of an extraordinary light beam (Poynting vector) does not coincide with its wave-vector direction. In order to avoid disturbing

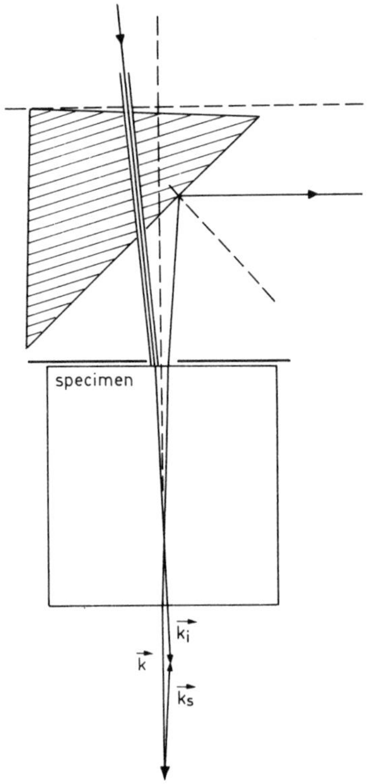

Fig.14 Experimental setup for the observation of backward scattering by phonons, from /76/.

reflections from the 'first' surface of the sample, the weakly focused incident laser beam has necessarily to be shielded by a narrow tube, as shown in Fig.14. A large angle between $\underline{k}_i$ and $\underline{k}_s$ magnifies the error of the phonon wave-vector direction in the crystal whereas a small angle makes experimental observation more difficult.

Experimental results on directional dispersion have become known during the last few years for α quartz /71/, $NaNO_2$ /93/, Be $SO_4 \cdot 4H_2O$ /94, 95/, $LiIO_3$ /96, 97/, Te /98/ and $LiNbO_3$ /99/. We furthermore refer to calculations for α quartz /135/ and $BaTiO_3$ /136/. Complete data concerning all the extraordinary branches of a uniaxial crystal have been presented in /99/. We point out that the frequencies of only-IR-active phonons for the principal directions may be extrapolated from the Raman data for directional dispersion. As an example we cite measurements on α quartz by Scott /71/. In this material the only-IR-active A_2 modes are coupled to E modes which are simultaneously IR- and Raman-active for arbitrary $\underline{k}$ directions.

Calculations of directional-dispersion branches in uniaxial crystals from (4-40) require determination of the LO and TO frequencies of all polar modes in the principal directions $\underline{k} \parallel z$ and $\underline{k} \perp z$. Such experiments have frequently been carried out by right-angle scattering on samples cut in such a way that the surfaces form angles of 45° with the optic axis. Suitable scattering geometries are illustrated in Fig.15. Note that for $\underline{k} \parallel z$ no E(LO) modes are allowed. Experiments of this kind were first published by Scott and Porto in 1967 /91/.

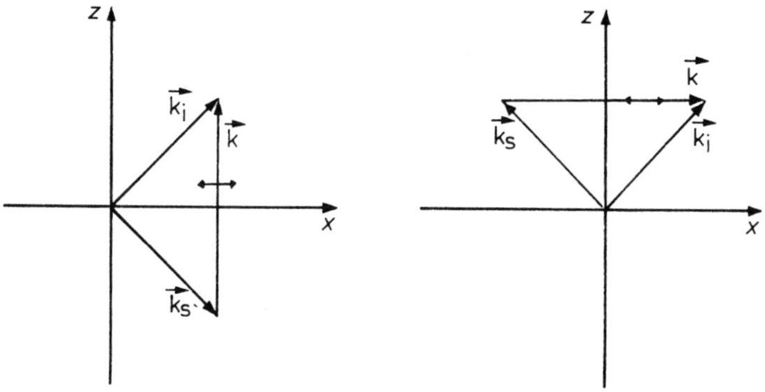

Fig.15 Scattering geometries allowing the determination of the LO frequencies of twofold degenerate modes. For wave-vector directions parallel to the optic axis no E(TO) modes are allowed, see /91/.

According to (4-14) and Fig.3, the frequencies of both A and E phonons have to form a sequence TO, LO, TO, LO,...,TO, LO from lower to higher wave numbers, starting with a transverse mode and ending with the highest-frequency LO mode. This is necessary because there is always a zero of the dielectric function between two poles of the Kurosawa relation. The coupling of phonons due to directional dispersion thus starts with the lowest-frequency TO mode coupled to the next phonon of symmetry A or E and type LO or TO at a higher frequency, and so on. The number of TO ↔ TO couplings thus has to be equal to the number of LO ↔ LO couplings. A look at these rules reveals an important source of error in interpreting spectra: a single second-order phonon erroneously assigned to first order, or a weak ghost due to reflections assigned as, for instance, an LO mode can disrupt the correct sequence and suggest incorrect couplings for all the following branches at higher frequencies, see /72, 99, 100/. Directional-dispersion measurements of extraordinary phonons therefore seem very useful for determining the correct assignment of the fundamental frequencies.

As an example, we reproduce in Fig.16 a directional-dispersion spectra series of $LiNbO_3$. The spectra for different values of θ were recorded on a series of samples with two parallel surfaces forming different angles with the optic axis. Five of the branches can easily be observed in this spectra series. $\theta = 0^{\circ}$ corresponds to $z(xx)\bar{z}$ and $\theta = \pi/2$ to $x(zz)\bar{x}$ scattering. The symbols used for the scattering geometries are those introduced by Porto, Giordmaine and Damen /285/. The letters outside the brackets indicate the direction of the incident laser beam to the left and the direction of the observed radiation to the right. The two letters in brackets denote the corresponding polarizations to left and right, respectively. These letters are identical with the subscripts of the Raman-tensor element. Thus, $\theta = 0^{\circ}$ corresponds to S_{xx}- and $\theta = \pi/2$ to S_{zz} scattering in Fig.16. In order to observe all 13 dispersion branches of the material, different spectra series starting with, for instance, $z(xy)\bar{z}$, $z(yx)\bar{z}$, or $z(yy)\bar{z}$ for $\theta = 0^{\circ}$ have to be recorded. Fig.17 is a comparison between the data points and branches calculated on the basis of (4-40).

Finally, it should be pointed out that a 'fitting' of one directional-dispersion branch will always influence the form of all the other branches because of the coupling described by (4-40). This can also be realized intuitively because the product of all TO-LO splittings determines the dielectric constants for the principal directions, see the LST relation in its generalized form (4-15). For arbitrary directions, on the other hand, the dielectric function is determined by (4-37). An error in the assignment is therefore possible only for branches with very small dispersion where the contribution to the dielectric function is negligibly small. Recent measurements of surface polaritons have shown that the almost constant branch 12 in Fig.17 should be located at $\sim$ 840 cm^{-1} whereas the Raman lines at $\sim$ 740 cm^{-1}, earlier assigned to be of first order /99, 100/, are due to a second-order process, see /197, 253/. This demonstrates the limit of the method described. In Fig.18 the assignments, wave numbers, and the principle of the couplings are schematically summarized in order to show the TO-LO sequences of A_1- and E-type modes in $LiNbO_3$.

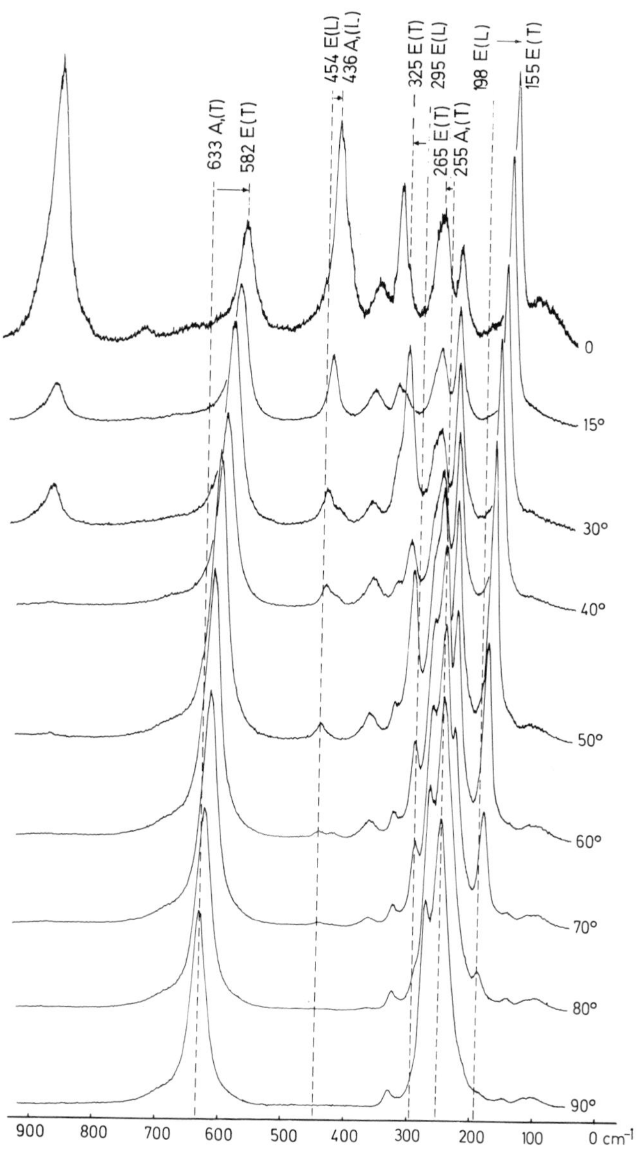

Fig.16 Experimental data on the directional dispersion of extraordinary phonons in $LiNbO_3$. The experiments were carried out by backward scattering, $\theta = 0^\circ$ corresponding to $z(x\ x)\bar{z}$ and $\theta = \pi/2$ corresponding to $x(z\ z)\bar{x}$ geometry, from /99/.

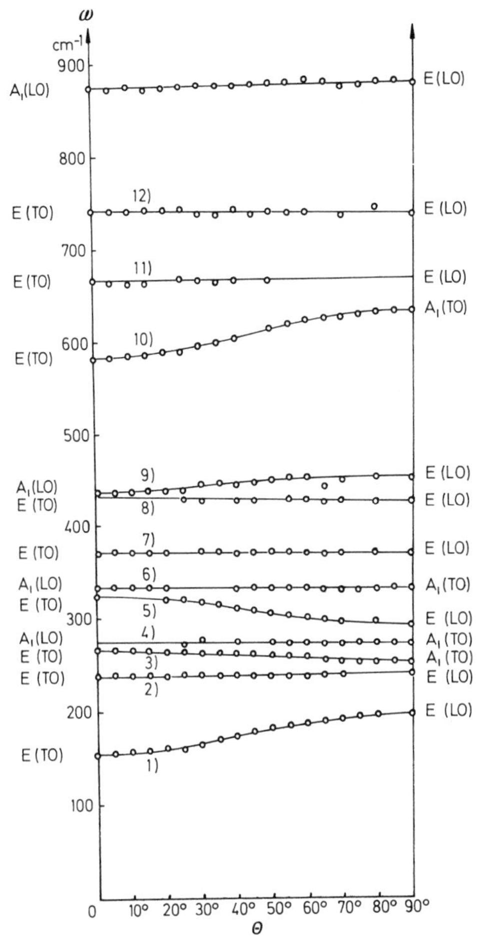

Fig.17 Calculated dispersion branches and experimental data points for the directional dispersion of extraordinary phonons in $LiNbO_3$, from /99/.

<u>Additional Literature</u>

Tsuboi, M., Terada, M. Kajiura, T.: Raman effect in ferroelectric sodium nitrite crystal. II. Dependence of B_2-type phonon spectrum on propagation direction /216/.

Anda, E.: Angular dispersion of large wave-vector polaritons in ferroelectric $NaNO_2$ /227/.

Shapiro, S.M., Axe, J.D.: Raman scattering from polar phonons /348/.

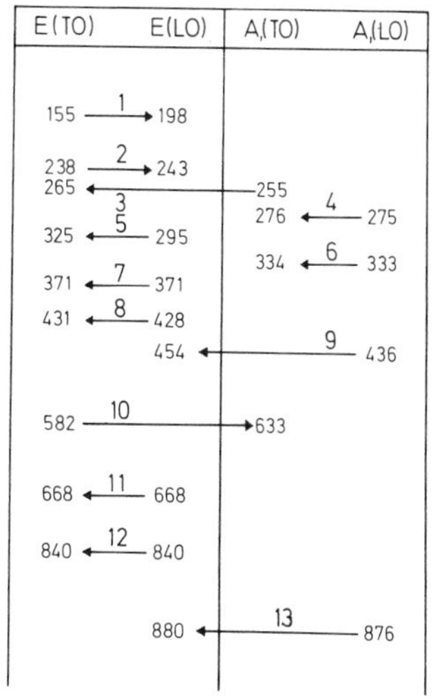

Fig.18 Assignments and wave numbers of the directional-dispersion branches in $LiNbO_3$ showing the TO-LO sequency of A_1 and E phonon modes, from /99/.

4.7 EXTRAORDINARY POLARITONS IN UNIAXIAL CRYSTALS (OBLIQUE POLARITONS)

For small wave-vectors $\underline{k} \to 0$ (long-wavelength limit) the dispersion relation of extraordinary polaritons (4-37) reduces to

$$\omega^2 \varepsilon_\perp(\omega)\varepsilon_\parallel(\omega) = 0 \quad . \tag{4-52}$$

The solutions of this equation correctly describe the region $k \ll 10^3\ \mathrm{cm}^{-1}$ of the 1BZ where the crystal lattice is completely coupled to the electromagnetic waves. The lowest frequency polariton branch ends at $\omega^2 = 0$, as in cubic crystals, see Figs.2 and 4, whereas the solutions obtained from $\varepsilon_\perp(\omega) = 0$ and $\varepsilon_\parallel(\omega) = 0$ coincide with those of the longitudinal phonons. $\varepsilon_\parallel(\omega) = 0$ leads to frequencies equal to those of the totally symmetric LO phonons and $\varepsilon_\perp(\omega) = 0$ to those of twofold-degenerate type. These solutions depend neither on the direction of the wave-vector nor on its magnitude. Thus all branches except the lowest are without dispersion in the region $k \ll 10^3\ \mathrm{cm}^{-1}$ of the 1BZ.

In the past, dispersion effects of extraordinary polaritons with $10^3 \lesssim k \lesssim 2 \times 10^4$ cm^{-1} have frequently been discussed for two limiting cases: (a) 'dominating anisotropy', with means small TO-LO splittings and large directional dispersion, and (b) 'dominating long-range electric fields' which implies large TO-LO splittings and almost vanishing directional dispersion (quasi-cubic crystals). These two limiting cases were first discussed by Merten in 1961 /65/ and later reviewed by Loudon /31/ and Claus /76/. Experimental work over the last few years, however, has shown that in most uniaxial polyatomic crystals of interest some modes will always show large TO-LO splittings and others predominately directional dispersion so that the 'materials' in general cannot simply be described by one of the limiting cases in question. The discussions cited are therefore only of interest in order to demonstrate some principles of the theory and can be excluded from our text without loss of information.

For every fixed wave-vector direction in uniaxial crystals the dispersion relation of the extraordinary polaritons (4-37) determines a complete set of dispersion curves $\omega^2 = \omega^2(k)$ in the polariton region. According to (4-52), the frequencies of the branches for $k = 0$ are identical for all the different sets: $\omega_o = 0$ and $\omega_i = \omega_{Li}$.

In the short-wavelength limit ($k \to \infty$), on the other hand, the mode frequencies of a certain set corresponding to a fixed angle $\theta = \theta'$ are equal to those derived from (4-39) or (4-40). For $LiNbO_3$ these can be seen directly in the ω,θ diagram reproduced in Fig.17. The frequencies $\omega(k)$ recorded for $k \simeq 10^3$ cm^{-1} and $k \approx 10^5$ cm^{-1} are connected by the polariton-dispersion curves. In Figs.19 to 23 sets of extraordinary polariton-dispersion branches of $LiNbO_3$ are reproduced for $\theta = 90^\circ$, 80°, 60°, 30° and 0°, respectively. All the diagrams clearly show identical mode frequencies for $k = 0$ whereas for $k = 1.5 \times 10^4$ cm^{-1} they vary with θ.

In order to demonstrate the main characteristics of extraordinary polaritons, we now discuss in detail the two limiting cases, $\theta = \pi/2$ and $\theta = 0$.

The general dispersion relation (4-37) can be written

$$(c^2k^2/\omega^2)(\varepsilon_\perp(\omega)\sin^2\theta+\varepsilon_\parallel(\omega)\cos^2\theta)-\varepsilon_\perp(\omega)\varepsilon_\parallel(\omega) = 0 \quad . \qquad (4\text{-}53)$$

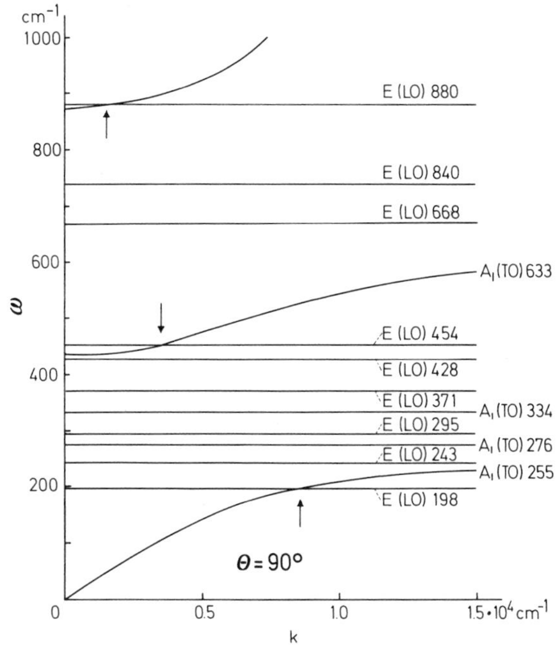

Fig.19 Extraordinary polariton-dispersion branches in $LiNbO_3$ for $\theta = \pi/2$. All E(LO) modes are without dispersion. Only the A_1(TO) branches depend on the magnitude of the wave-vector, from /103/.

It follows directly that

$$\varepsilon_\perp(\omega)\left[(c^2k^2/\omega^2)-\varepsilon_\parallel(\omega)\right] = 0 \quad \text{for} \quad \theta = \pi/2 \quad , \tag{4-54}$$

and

$$\varepsilon_\parallel(\omega)\left[(c^2k^2/\omega^2)-\varepsilon_\perp(\omega)\right] = 0 \quad \text{for} \quad \theta = 0 \quad . \tag{4-55}$$

The dispersion relation can thus be factorized in these two cases and accordingly the solutions become of two different types in both cases.

a) For $\theta = \pi/2$, the zeros of $\varepsilon_\perp(\omega) = 0$ determine the E(LO) phonons, which do not depend on the magnitude of the wave-vector, while the zeros of the bracket describe purely transverse polaritons associated with phonons of the A type. Only one of the dielectric functions, $\varepsilon_\parallel(\omega)$, is involved.

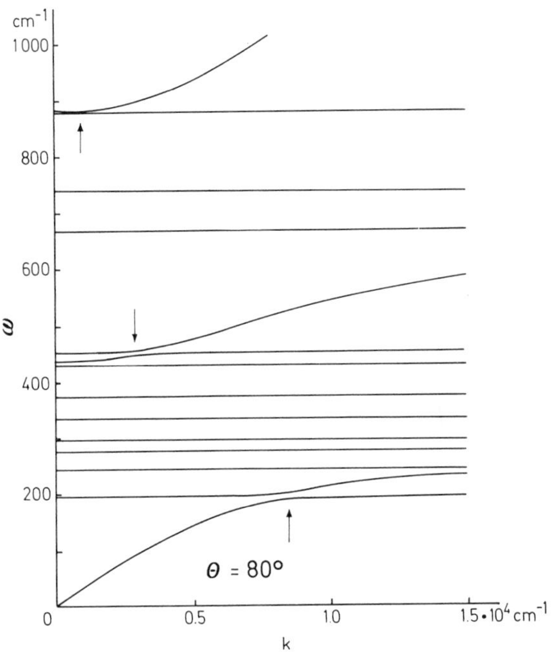

Fig.20 Extraordinary polariton branches in $LiNbO_3$ for $\theta = 80^{\circ}$, see /103/.

b) For $\theta = 0^{\circ}$, the zeros of $\omega_{\|}(\omega) = 0$ determine the A(LO) modes and the zeros of the bracket describe purely transverse polaritons associated with the E(TO) phonons. In this case, too, both types of solutions are linearly independent. We especially note that the extraordinary polariton-dispersion branches for $\theta = 0^{\circ}$ coincide with those of the directionally independent ordinary polaritons. The zeros of the bracket in (4-55) are identical to those of the first one in (4-36). This can easily be verified by comparing Figs.19 and 23.

The linear independence of the eigenvectors of purely transverse polaritons and LO phonons for the principal directions allows a crossing of dispersion branches to take place. The modes are decoupled. $LiNbO_3$ shows four crossings of this type for $\theta = 90^{\circ}$. Three of them have been indicated by arrows in Fig.19. The fourth one appears to the right of the figure between the E(LO) phonon at 243 cm^{-1} and the A_1(TO) polariton mode moving towards 255 cm^{-1} for large k values.

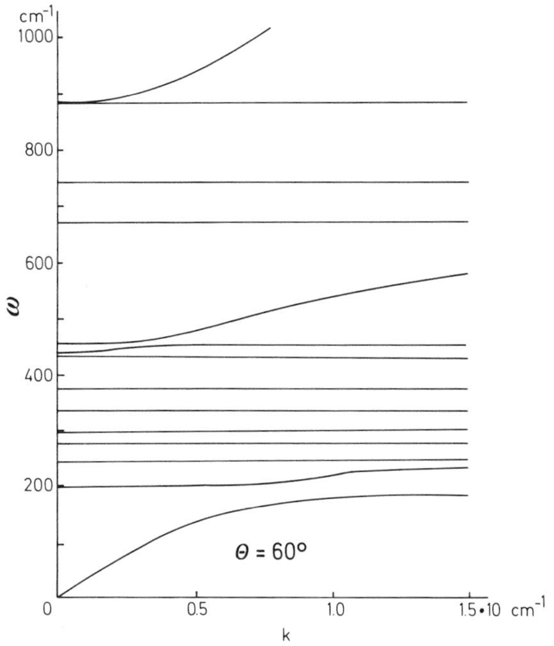

Fig.21 Extraordinary polariton branches in $LiNbO_3$ for $\theta = 60^\circ$, see /103/.

According to (4-37) or (4-53), there are no modes of either exactly A or exactly E type for arbitrary wave-vector directions because the dispersion relation can no longer be factorized. Both dielectric functions, $\varepsilon_{\parallel}(\omega)$ and $\varepsilon_{\perp}(\omega)$, are always involved. All dispersion branches then describe modes of mixed type A + E and TO + LO in the same way as the extraordinary phonons discussed in 4.6 were mixed for arbitrary wave-vector directions. This means that the crossing points are lifted. The modes become coupled for general wave-vector directions, see Fig.20 (arrows). The interpretation of the two lowest-frequency branches in this figure is as follows: the quasi-A_1 (TO) polariton couples with the quasi-E(LO) phonon for decreasing wave-vectors and ends at 198 cm^{-1} for k = 0. Correspondingly, the quasi-E(LO) phonon for smaller wave-vectors changes to quasi-A_1(TO) type and shows polariton character with a mode frequency decreasing to $\omega = 0$ for k = 0.

Mathematically, we can regard the crossing points appearing for $k \perp z$ and $k \parallel z$ as singularities. The polarizations of the modes involved are changed discontinuously from A(TO) to E(LO) and vice

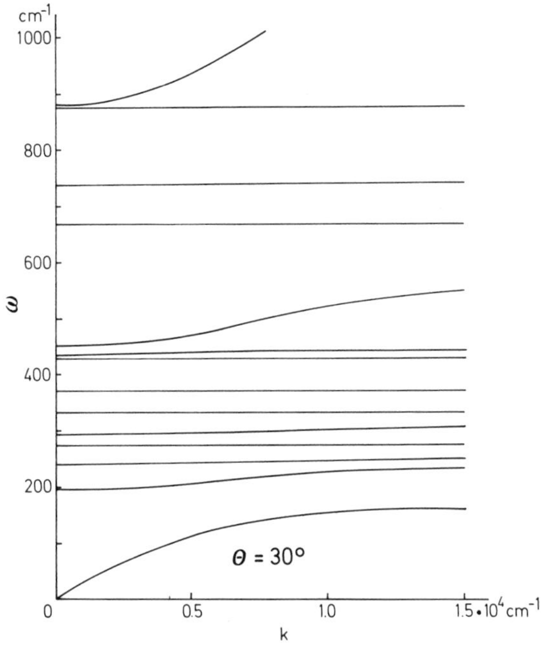

Fig.22 Extraordinary polariton branches in $LiNbO_3$ for $\theta = 30^{\circ}$, see /103/.

versa, or from E(TO) to A(LO) and vice versa. The latter situation appears for k ∥ z. For decreasing angles θ the coupling of the modes becomes stronger (illustrated in Figs.21 and 22) and accordingly the frequency splitting at the resonance points rises until finally we have, for $\theta = 0^{\circ}$, the second limiting case (b) with pure mode symmetries and exact polarizations (Fig.23). Another example is shown in Fig.24, where the calculated extraordinary polariton branches of Te have been sketched for $\theta = 90^{\circ}$, 80° and 60°, /74/. The lifting of two crossing points appearing for $\theta = 90^{\circ}$ can be very clearly observed, see also corresponding calculations for α quartz /134/.

The 'crossing phenomenon' has been studied theoretically in detail by Lamprecht and Merten /68/, neglecting damping of the modes, although it also holds for the case of small damping /70/. For large damping, crossing points between transverse modes which are not lifted for off-principal directions have been predicted by Borstel and Merten /101/.

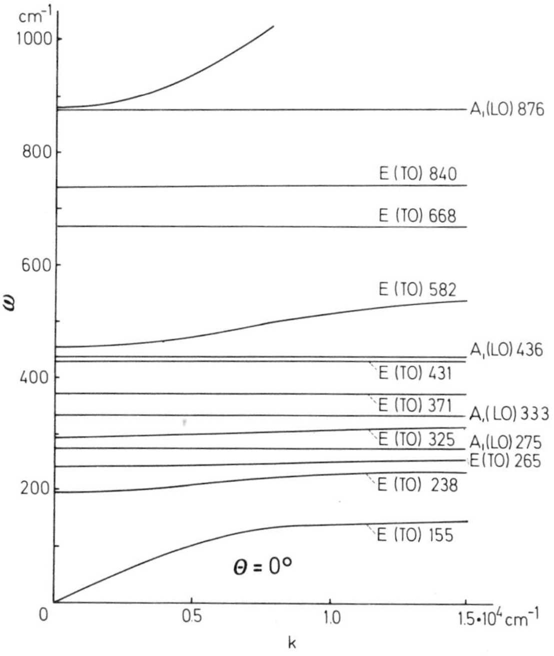

Fig.23 Extraordinary polariton-dispersion branches in $LiNbO_3$ for $\theta = 0^o$. All A_1(LO) modes are without dispersion. Only the E(TO) branches depend on the magnitude of the wave-vector, from /103/.

The appearance and lifting of TO-LO crossing points seems to be the most characteristic feature of extraordinary polaritons. In fact, this phenomenon demonstrates the mode-mixing and the directional dispersion as well as the appearance of modes with pure symmetries and exact polarizations for the principal directions.

Detailed experimental studies concerning the crossing point between the lowest frequency A_1(TO) polariton mode and the E(LO) mode at 198 cm^{-1} in $LiNbO_3$ have been published /102/ and /103/. According to Fig.9 the dispersion of extraordinary polaritons can be recorded as a function of different fixed pairs of k and θ simply by varying the scattering angle ϕ. The observation of extraordinary polaritons with k and θ varying independently has been achieved by the use of different crystal samples, their optic axes forming different angles with two parallel surfaces. In Fig.9 this implies that the optic axis forms different angles with the dashed line. For a comparison of calculated curves and experimental data, see Figs. 25

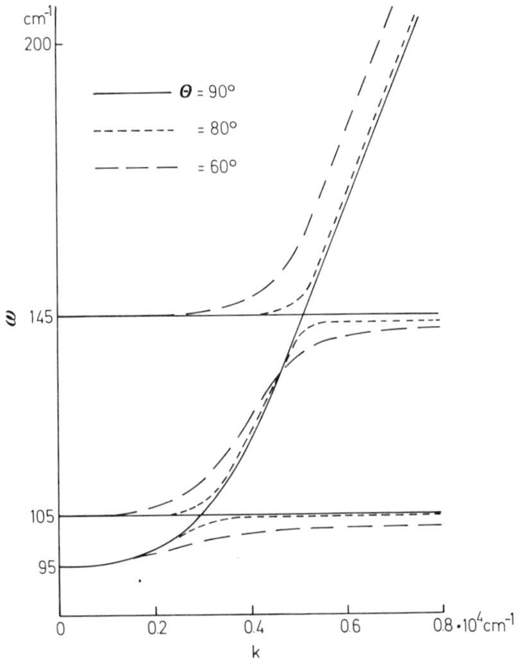

Fig.24 'Crossing points' appearing for $\theta = 90^{\circ}$ in Te, from /74/.

and 26. Data points for the horizontal dispersion 'curve' of the E(LO) phonon have not been plotted in Fig.25 because the observation of this phonon is experimentally trivial for all k in this region. Three corresponding spectra series are reproduced in Fig.27. In Fig.27a $\theta = \pi/2$ for all scattering angles ϕ. $\phi = 0$ corresponds to the geometry x(zz)x, whereas increasing ϕ yields x(zz)x + y(zz)x. The modes observed are therefore purely of symmetry A_1. In Figs. 27b and 27c $\theta = 87^{\circ}$ and $\theta = 82^{\circ}$ at approximately maximum resonance; the increasing frequency splitting with decreasing θ can easily be observed.

The rapid exchange of the scattering intensity of the polaritons is caused mainly by the change of symmetry. The polariton branch of quasi A_1(TO) type changes to quasi E(LO) type and vice versa, see above. Owing to the oblique orientation of the optic axis, the straight forward scattering in Figs.27b and c ($\phi = 0$) corresponds to x(zz)x mixed with some $\bar{z}(xx)\bar{z}$ scattering. As can be seen from the Raman tensors for the space group C_{3v}^{6}(= R3c) of $LiNbO_3$ (Appdx 4)

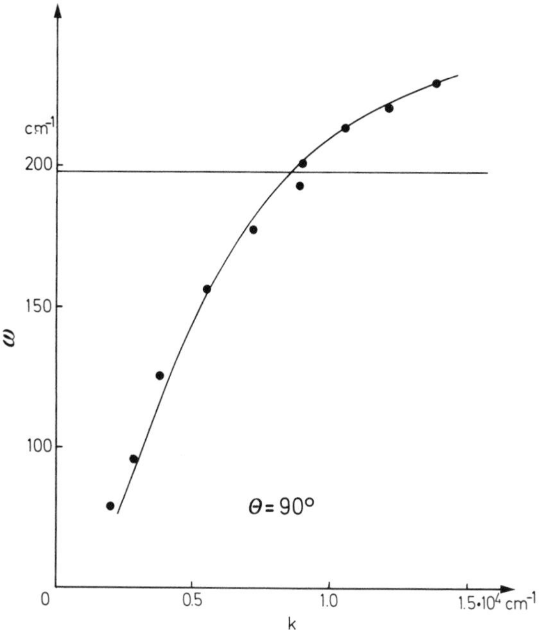

Fig.25 Experimental data points and calculated dispersion curves concerning the crossing point between the E(LO) branch at 198 cm^{-1} and the polariton branch associated with the A_1(TO) phonon at 253 cm^{-1} for $\theta = \pi/2$ in $LiNbO_3$, from /103/.

this causes modes of twofold-degenerate type to be observed simultaneously with those of totally symmetric type. As a result the quasi-E(LO) phonon appears stronger for $\phi = 1.4^{\circ}$ and $\phi = 3.6^{\circ}$ in Fig.27c than in the corresponding spectra of Fig.27b. In Fig.27a, on the other hand, this phonon cannot be observed at all for the scattering angles in question. The increasing intensity is thus not caused only by the mixed symmetry ('A' + E) of the 'quasi'-E(LO) phonon.

When the Raman tensor elements of polar modes do not vanish, we can in principle distinguish four different cases allowing the observation of polaritons in uniaxial crystals.

1) Ordinary incident and ordinary scattered photons

$n_i = n_o(\omega_i)$ and $n_s = n_o(\omega_s)$

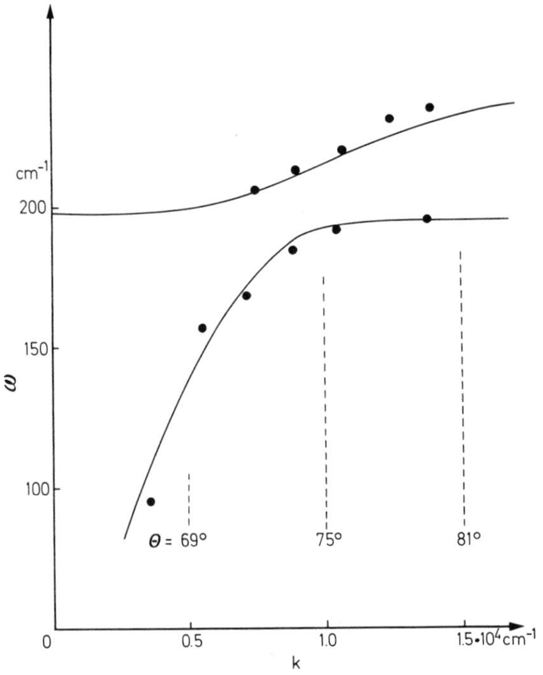

Fig.26 Experimental data points and calculated curves concerning the lifted crossing point, Fig.25, from /102, 103/.

2) Ordinary incident and extraordinary scattered photons

$$n_i = n_o(\omega_i) \quad \text{and} \quad n_s = n_e(\omega_s) \quad ,$$

3) Extraordinary incident and extraordinary scattered photons

$$n_i = n_e(\omega_i) \quad \text{and} \quad n_s = n_e(\omega_s) \quad ,$$

4) Extraordinary incident and ordinary scattered photons

$$n_i = n_e(\omega_i) \quad \text{and} \quad n_s = n_o(\omega_s) \quad .$$

n_o and n_e denote the ordinary and extraordinary refractive indices of the medium, respectively.

Among the nineteen classes of uniaxial crystals there are six with an inversion center. This causes IR-active modes to be Raman-inac-

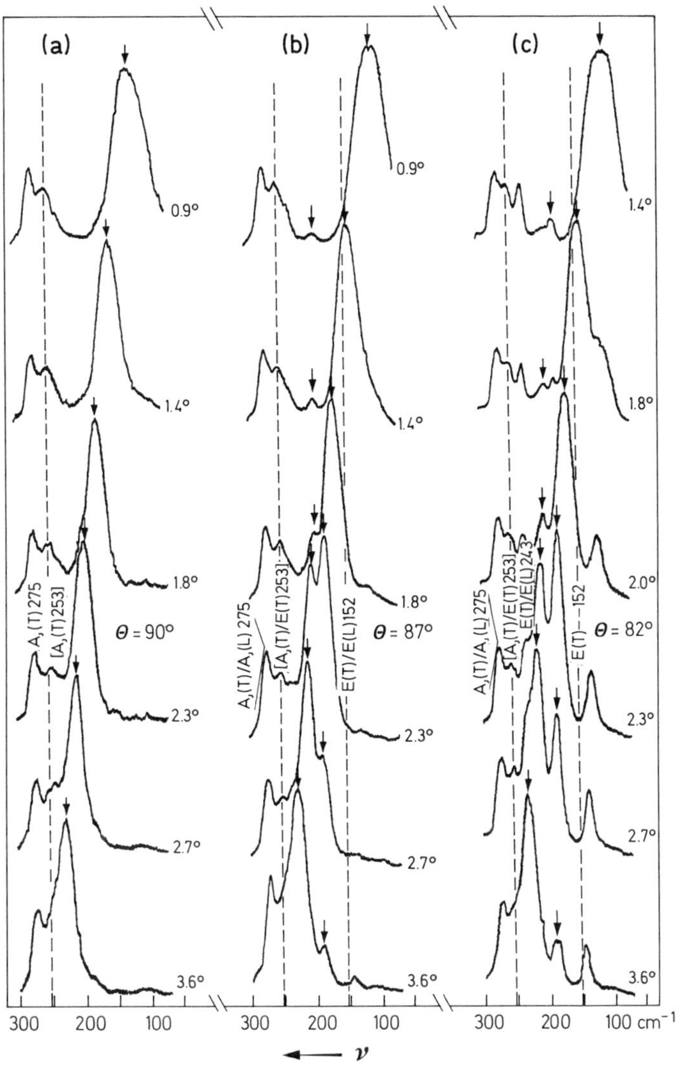

Fig.27 a) Spectral series showing the crossing point between the E(LO) branch at 198 cm^{-1} and the polariton branch associated with the A_1(TO) phonon at 253 cm^{-1} for $\theta = \pi/2$ in $LiNbO_3$. b) and c) Spectra series showing the lifted crossing point for general wave-vector directions. The spectral series c) furthermore allows simultaneous observation of the lowest-frequency ordinary and extraordinary polariton branches moving towards $\omega = 0$ for $k \to 0$, from /102, 103/.

tive and vice versa, see 2.3 to 2.6. In materials belonging to one of the thirteen remaining crystal classes polaritons may be directly observed by light scattering. These classes are $\mathcal{C}_3$ (= 3), $\mathcal{D}_3$ (= 32), $\mathcal{C}_{3v}$ (= 3m), $\mathcal{C}_4$ (= 4), S_4 (= $\bar{4}$), $\mathcal{C}_{4v}$ (= 4mm), $\mathcal{D}_4$ (= 422), $\mathcal{D}_{2d}$ (= $\bar{4}$2m), $\mathcal{C}_6$ (= 6), $\mathcal{C}_{3h}$ (= $\bar{6}$), $\mathcal{D}_6$ (= 622), $\mathcal{C}_{6v}$ (= 6mm), and $\mathcal{D}_{3h}$ (= $\bar{6}$m2).

If a scattering geometry corresponding to case 1) described above is chosen, the situation is analogous to that in cubic crystals, see 4.3. The ordinary refractive index does not depend on the wave-vector direction. Only the Raman tensor elements S_{xx} and S_{yy} are involved. In the crystal classes $\mathcal{C}_3$ (= 3) and $\mathcal{C}_{3v}$ (= 3m) polaritons associated with totally symmetric phonons as well as with phonons to twofold-degenerate type may be observed simultaneously (N.B. simultaneous observation of extraordinary and ordinary polaritons).

Case 3) is also similar to the situation for cubic crystals. The tensor element S_{zz} involved vanishes for all vibrational species except the totally symmetric modes. In contrast to case 1), however, we have to examine whether the directional dependence of the extraordinary refractive index can be neglected or not. Moreover, the rotation of the polarization plane in optically active media may simulate a breakdown of the selection rules. The optically active crystal classes are indicated in Appdx 4. In the case of straight forward scattering along the x axis, for instance, the approximation $n_i \approx n_s$ neglects only the dispersion $n = n(\omega)$. For larger scattering angles ϕ, however, this applies only when both $\underline{k}_i$ and $\underline{k}_s$ propagate perpendicular to the optic axis (z) with all scattering triangles $\underline{k}_i = \underline{k}_s + \underline{k}$ lying in the optically isotropic plane (xy).

Similar features have to be taken into account in cases 2) and 4) when $\underline{k}_i$ and $\underline{k}_s$ are not perpendicular to the z axis. Experiments due to cases 2) and 4) are distinguished from the others by the fact that the refractive indices n_i and n_s can be regarded as approximately equal only for vanishing birefringence. In general, however, this is not true.

According to (4-28) the magnitude of the polariton wave-vector for a Stokes process $\underline{k}_i = \underline{k}_s + \underline{k}$ is

$$k = 2\pi\left[(n_i\nu_i)^2+(n_s\nu_s)^2-2n_in_s\nu_i\nu_s \cos\phi\right]^{1/2} \quad , \tag{4-56}$$

where $k_i = 2\pi n_i \nu_i$ and $k_s = 2\pi n_s \nu_s$. n_ρ (ρ = i, s) are the refractive indices for the incident and scattered photons. $\nu_\rho = 1/\lambda_\rho$ are the wave numbers. λ_ρ again denotes the wavelength in vacuum. Written in parameter form, (4-56) becomes

$$k = 2\pi\left[n_s^2\nu^2+2n_s\nu_i(n_i \cos\phi-n_s)\nu+\nu_i^2(n_i^2+n_s^2-2n_in_s \cos\phi)\right]^{1/2} \quad . \quad (4\text{-}57)$$

This is an equation for hyperbolas in the ν,k diagram. The parameter is ϕ (note that in the derivation of (4-57) energy conservation requires $\omega_i = \omega_s+\omega$).

In the limiting case $\phi = 0$ we obtain the two linear equations for the asymptotes. They determine the maximum polariton frequency shifts observable for a certain scattering geometry, as described earlier in 4.3. The equations are

$$\nu = (1/2\pi n_s)k + (\nu_i/n_s)(n_s - n_i) \quad (4\text{-}58)$$

and

$$\nu = -(1/2\pi n_s)k + (\nu_i/n_s)(n_s - n_i) \quad . \quad (4\text{-}59)$$

In Fig.28 the ordinary and extraordinary polariton dispersion branches of a uniaxial crystal with two atoms in the unit cell have been sketched for wave-vector directions perpendicular to the optic axis. Fig.28b shows the two limiting lines described by (4-58) and (4-59). They have been indicated by $\phi = 0$. The intersection with the ω axis is proportional to the birefringence of the crystal. The wave number of the intersection point is

$$\nu' = \nu_i(n_s - n_i)/n_s \quad . \quad (4\text{-}60)$$

The situation for $\nu' = 0$, illustrated in Fig.28a, appears for $n_i = n_s$, corresponding to cases 1) and 3) described above, and will be of importance mainly for the observation of polaritons associated with phonons of totally symmetric type. On the other hand, $\nu' \neq 0$ (Fig.28b) normally holds when recording polaritons associated with phonons of E type. For $\phi > 0$, (4-57) determines the hyperbolas sketched in the figures. Their points of intersection with the dispersion curves again determine which polariton frequencies are observed. The dispersion branches are described as $\omega^2 = \omega^2(k)$

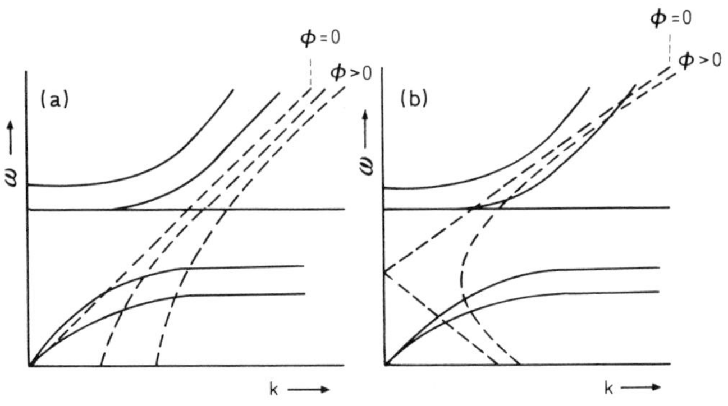

Fig.28 Observation of polaritons in uniaxial crystal: a) light scattering by diagonal elements of the Raman tensor, for instance, S_{zz} scattering corresponding to A-type polaritons; b) light scattering by off-diagonal elements corresponding to, for instance, E-type polaritons, from /76/.

by, for instance (4-36), while (4-56) determines $k = k(\phi)$. Solving (4-36) and (4-56) simultaneously, we obtain $\omega^2 = \omega^2(\phi)$.

Among the thirteen classes of uniaxial crystals allowing Raman scattering by polar modes there are only five in which modes of twofold degenerate type can be observed by light scattering due to diagonal elements of the Raman tensor. These crystal classes are $\mathcal{C}_3$ (= 3), $\mathcal{D}_3$ (= 32), $\mathcal{C}_{3v}$ (= 3m), $\mathcal{C}_{3h}$ (= $\bar{6}$), and $\mathcal{D}_{3h}$ (= $\bar{6}$m2). Certain off-diagonal elements, on the other hand, will always allow scattering by E polaritons. In the two crystal classes $\mathcal{C}_{3h}$ (= $\bar{6}$) and $\mathcal{D}_{3h}$ (= $\bar{6}$m2), these are the elements S_{xy} and S_{yx} corresponding to case 1) and Fig.28a. In the eleven remaining classes, on the other hand, the tensor elements S_{xz}, S_{zx}, S_{yz} and S_{zy} are generally involved. The experiments have to be carried out in the way described in 2) and 4) above, Fig.28b.

Further experimental techniques can be applied to uniaxial crystals when making use of birefringence effects. This was first demonstrated by Porto, Tell and Damen on ZnO /104/. In spite of the symmetry of the Raman tensor ($S_{xz} = S_{zx}$ or $S_{yz} = S_{zy}$), scattering experiments corresponding to cases 2) and 4) in this material will achieve different maximum frequency shifts. From (4-60) it can be seen that the limiting lines for ZnO have a positive intersection with the ω axis in case 2) and a negative one of the same magnitude when changing the geometry to case 4). This happens because ZnO is an optically positive crystal where $n_e > n_o$. In Fig.29 the phonon-like polariton-dispersion curve of ZnO is plotted together with the limiting lines and the hyperbolas for different values of the parameter ϕ in both cases. Obviously S_{yz} scattering, case 2), allows the observation of larger frequency shifts. Regarding the wave-vector triangles it will readily be realized that for the same scattering angle ϕ different polariton wave-vectors are involved: the magnitudes of the incident and scattered photon wave-vectors $k_\rho = 2\pi n_\rho/\lambda_\rho$ are determined in two different ways by the refractive indices. The phenomenon has been illustrated for $n_o > n_e$ in Fig.30, corresponding to analogous experiments in $LiIO_3$, which is an optically negative crystal, see /105/. For suitable scattering geometries with $n_s > n_i$ the magnitude of the scattered photon wave-vector $\underline{k}_s$ may be even larger than $|\underline{k}_i|$. The use of birefringence effects in the way described will in principle allow Raman-scattering experiments exactly in the center of the 1BZ.

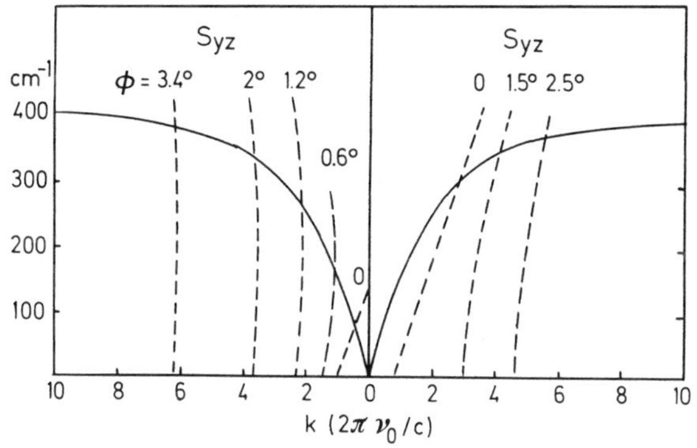

Fig.29 Phonon-polariton dispersion curve of ZnO. Limiting lines and hyperbolas according to (4-58), (4-59) and (4-60) are sketched for S_{yz} and S_{zy} scattering, from /104/.

α quartz was the first polyatomic uniaxial crystal in which polariton scattering was observed /106/ (Scott et al. in 1967). The dispersion branches of the eight transverse polariton modes of E type were calculated from a generalized dispersion relation published by Barker in 1964 /63/. This relation, however, holds only for the principal directions and the experimental work is accordingly concerned only with ordinary polaritons. Pinczuk, Burstein and Ushioda examined ordinary polariton branches in the tetragonal phase of $BaTiO_3$ (C_{4v}) (= 4mm) in the same way in 1969 /107/.

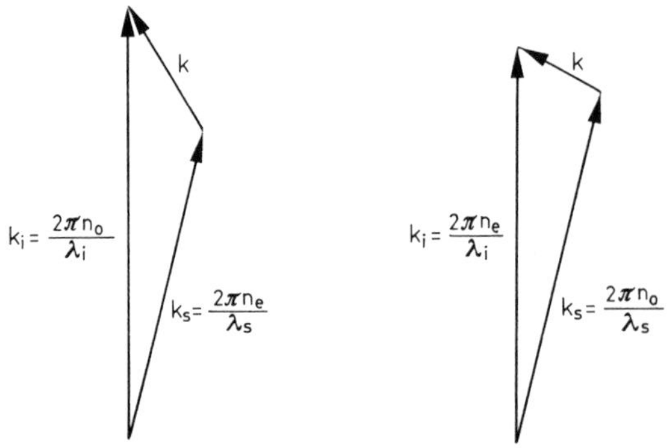

Fig.30 Dependence of polariton wave-vectors on the refractive indices of the incident and scattered photons in uniaxial crystals, from /105/.

Additional Literature

Khashkhozhev, Z.M., Lemanov, V.V., Pisarev, R.V.: Scattering of light by acoustic phonons and polaritons in $LiTaO_3$ /222/.

Burns, G.: Polariton results in ferroelectrics with the tungsten bronze crystal structure /232/.

Otaguro, W.S., Wiener-Avnear, E., Porto, S.P.S., Smit, J.: Oblique polaritons in uniaxial crystals: application to $LiIO_3$ /242/.

Hizhnyakov, V.V.: Effect of the polariton-phonon coupling on the dispersion /254/.

Laughman, L., Davis, L.W.: Raman scattering by pure and mixed polaritons in BeO /329/.

4.8 ORDINARY POLARITONS AND PARAMETRIC LUMINESCENCE IN UNIAXIAL CRYSTALS

Sections 4.6 and 4.7 contained a detailed discussion of the extraordinary modes. All derivations were based on (4-37) which in turn is derived from the zeros of the second bracket of the general polariton-dispersion relation in uniaxial crystals, (4-36). We now turn our attention to the zeros of the first bracket in this equation leading to the description of ordinary polaritons

$$(c^2k^2/\omega^2) = \varepsilon_\perp(\omega) \quad . \tag{4-61}$$

We already pointed out some characteristics in 4.4: these modes do not depend on the direction of the wave-vector because no components of the wave normal $\underline{s}$ appear in (4-61). Ordinary polariton modes thus show only dispersion due to the magnitude of $\underline{k}$ and the branches are consequently identical for all wave-vector directions in the crystal, see also Table 6. Furthermore, only one dielectric function $\varepsilon_{\perp}(\omega)$ is involved in (4-61). This implies that all ordinary polaritons are exactly of E type. Finally the modes are purely transverse for all wave-vector directions, which can be shown in the same way as was done for cubic crystals in 4.3. The proof is formally identical to the calculation in 4.3 because Huang's equations for ordinary polaritons in uniaxial crystals are identical to the set for cubic crystals.

In 4.7 we have shown that extraordinary polaritons degenerate with ordinary ones for $\underline{k} \parallel z$. The only difference between (4-55) and (4-61) is that no factor $\varepsilon_{\parallel}(\omega)$ determining exactly longitudinal A modes for $\varepsilon_{\parallel}(\omega) = 0$ appears in (4-61). Fig.23 therefore simultaneously illustrates the ordinary dispersion curves of $LiNbO_3$ by omitting the four A_1(LO) branches. For a better survey, the ten ordinary branches have been plotted separately in Fig.31. In the center of the 1BZ the frequencies coincide with those of the E(LO) modes: $\varepsilon_{\perp}(\omega) = 0$, see (4-61). For large k, on the other hand, they become equal to those of the E(TO) phonons. The corresponding wave numbers are given to left and right of the figure.

The dispersion branch ending at the highest E(LO) phonon frequency in the center of the 1BZ describes ordinary photons $k \to \infty$. The directional independence of ordinary polaritons corresponds to the fact that the ordinary refractive index known from crystal optics does not depend on the photon wave-vector direction. Similarly, the different slopes of the photon-like branches of extraordinary polaritons for different θ determine the varying phase velocities of the extraordinary ray for different $\underline{k}$ directions, see Figs.19 to 23.

The degeneration of extraordinary and ordinary polaritons for $\theta = 0^{\circ}$ may be confused with the fact that the ordinary modes are observed advantageously in experiments only for wave-vectors lying in the optically isotropic plane, which means $\theta = \pi/2$. We therefore remember that ordinary modes are exactly of type E(TO) for <u>all</u> wave-vector directions whereas extraordinary modes are of type

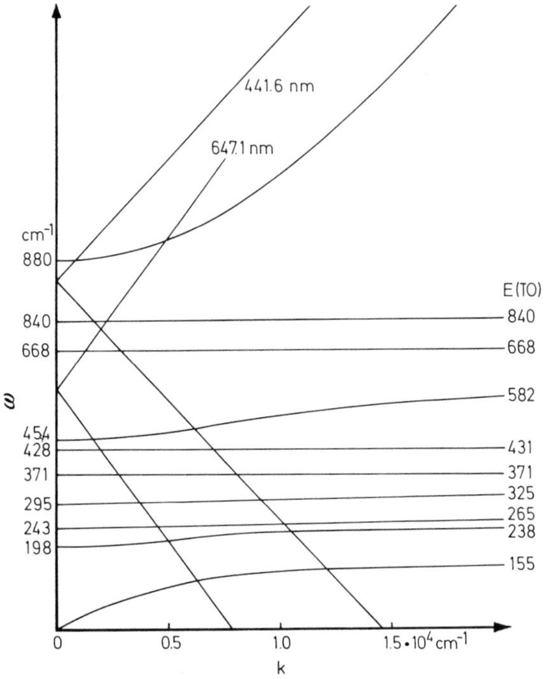

Fig.31 Ordinary polariton dispersion branches of $LiNbO_3$ (for arbitrary wave-vector directions). The limiting lines as determined by (4-58) and (4-59) have been plotted for krypton-laser excitation (647.1 nm) and for He-Cd-laser excitation (441.6 nm). The wave numbers on the right are those of the E(TO) phonons for $k \to \infty$, from /108/.

A(TO) or E(LO) only for $k \perp z$. Experiments allowing separate observation of E(TO) polaritons can thus be attained only by using scattering geometries with wave-vector triangles lying in the optically isotropic plane. The scattering angle ϕ can then be changed arbitrarily because the condition $k \perp z$ is obviously fulfilled for all ϕ. A(TO) modes are easily eliminated from the spectra by suitable polarizations of the laser beam and the observed scattered radiation so that the E(TO) spectra are not mixed up with Raman lines originating from extraordinary modes. Such scattering geometries are $\underline{k}_i(z,\ x \leftrightarrow y)\underline{k}_s$, $\underline{k}_i(y \leftrightarrow x,\ z)\underline{k}_s$, or $\underline{k}_i(y \leftrightarrow x,\ x \leftrightarrow y)\underline{k}_s$ for varying angles between $\underline{k}_i$ and $\underline{k}_s$. Extraordinary A(TO) polaritons may be separately recorded using geometries $\underline{k}_i(z\ z)\underline{k}_s$, with $\underline{k}_i$ and $\underline{k}_s$ forming an arbitrary angle ϕ in the optically isotropic plane.

Because of the dependence of the refractive indices on the frequency of the incident laser radiation, it is also the case in uniaxial crystals that the limiting lines determining the greatest polariton-frequency shifts observed can be rather different. In Fig.31 the situation has been sketched for He-Cd-laser excitation (441.6 nm) and Krypton-laser excitation (647.1 nm). The intersection points with the ω axis are derived from (4-60) and the slopes of the lines are $\pm 1/2\pi n_s$, see (4-58) and (4-59). Obviously the slopes are constant only when the refractive indices can be regarded as constant over the whole range of the Raman spectrum

$$n_s(\omega_i) \approx n_s(\omega') \quad . \tag{4-62}$$

ω' is the varying polariton frequency (Stokes or Anti-Stokes wave) and ω_i the frequency of the incident laser radiation as before. Only on this condition are the 'limiting lines' really 'lines'. In general, the approximation (4-62) is rather good in the region of phonon-like polaritons located a few hundred cm^{-1} away from the exciting line. According to Fig.28b, however, the photon-like polariton branches with large dispersion may also be observed by forward scattering. Experiments have shown that the approximation (4-62) usually cannot then be made because frequency shifts of several thousand wave numbers may easily be observed.

On comparing the limiting lines in Fig.31, it becomes obvious that Krypton-laser excitation is more suitable for investigations in the lower frequency region of $LiNbO_3$ whereas the He-Cd laser allows better observation of photon-like polaritons. A spectra series showing the lowest-frequency ordinary branch in this material is reproduced in Fig.32. Owing to reflections of the incident laser radiation on the second surface inside the sample and those of scattered radiation on the first surface, the Raman line at 154 cm^{-1} corresponding to the E(TO) phonon for $k \to \infty$ appears simultaneously with the polariton mode in all the scans. The maximum frequency shift, from 154 to 117 cm^{-1} for $\phi = 0$, is determined by the left intersection point on the corresponding branch in Fig.31.

In order to observe light-scattering processes originating from photon-like polaritons, a photographic method first described by Akhmanov et al. /109/ has proved very useful. This method has also been useful for recording phonon-like polaritons with small cross-sections, as we shall show. The principle of the experimental setup

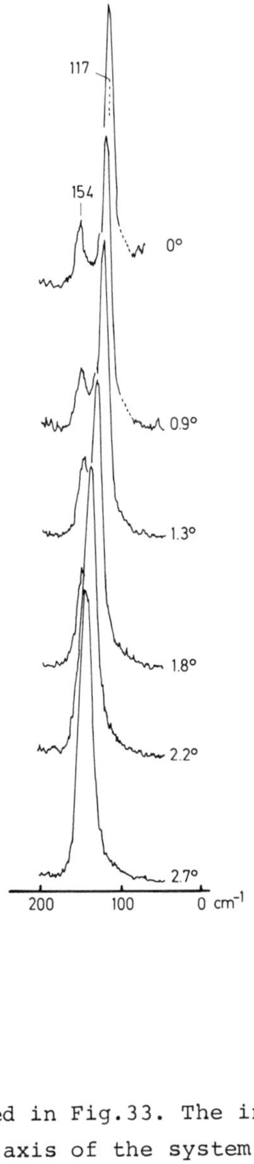

Fig.32 Raman-spectra series showing the lowest-frequency ordinary polariton branch of $LiNbO_3$, λ_i = 647.1 nm, from /108/.

is sketched in Fig.33. The incident laser beam propagates along the optic axis of the system (horizontal dashed line) toward the left of the figure. The luminescing volume is projected onto the entrance slit S of a spectrograph with the lens L. This lens is placed at focal distance from S so that all scattered radiation corresponding to a certain external angle ϕ' will be focused onto one point of the slit. (The internal scattering angle ϕ of course is smaller than ϕ'.) Radiation focused on the center of the slit originates from straight forward scattering, corresponding to pola-

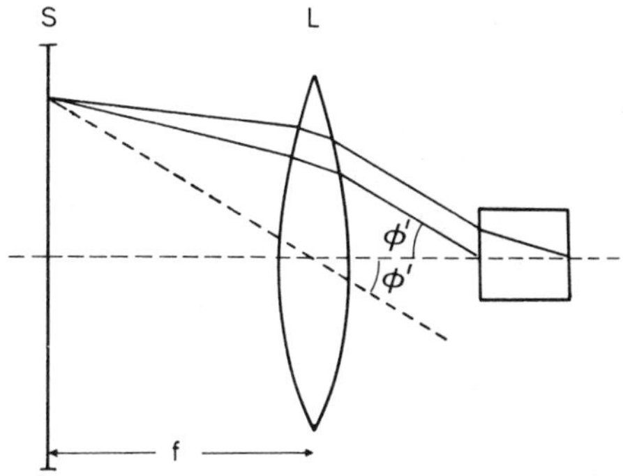

Fig.33 Experimental setup for photographic recording of polaritons, see /109/ and /108/. For explanation, see the text.

ritons with maximum frequency shifts towards smaller wave numbers. Positions on the slit at different distances symmetrically from its center thus correspond to different fixed external scattering angles ϕ' and $-\phi'$. In order to reduce stray light caused by the laser radiation in the spectrograph, two tricks may be used: (a) the incident laser beam may be focused just beside the center of the slit opening, so preventing the radiation from directly entering the spectrograph; (b) a filter may be placed behind the slit (inside the spectrograph). A typical spectrum, reproduced in Fig.34, shows the purely transverse phonon-like polariton associated with the A_1(TO) mode at 633 cm^{-1} in $LiNbO_3$. The spectrogram has been exposed in the way described. Due to some spillover scattering, the position of the phonon is again recorded as a weak spectral line, see Fig.32. The largest polariton shifts are observed in the center of the picture, corresponding to the center of the entrance slit, whereas the upper and lower parts of the figure corresponding to larger ϕ show the polariton moving asymptotically towards the A_1(TO)-phonon frequency.

Winter and Claus /108/ used this technique to record the polariton dispersion of the ordinary E(TO) branch in $LiNbO_3$ moving from 582 to 454 cm^{-1} for decreasing wave-vectors. Although this branch shows a rather strong dispersion, see Fig.31, the scattering cross-sections are extremely small over the whole k range in question. Backward

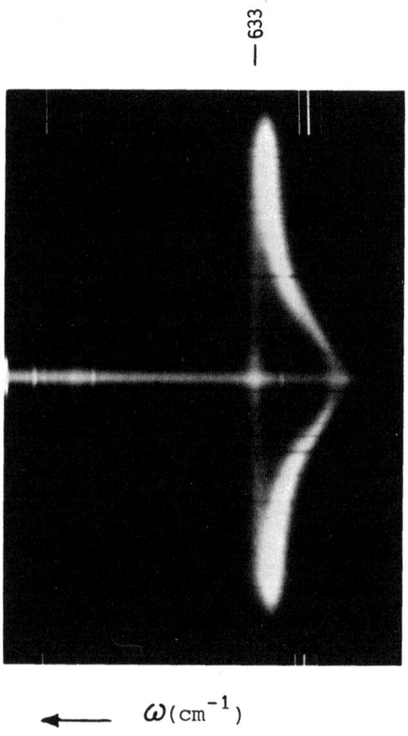

Fig.34

Spectrogram of polaritons associated with the A_1(TO) phonon at 633 cm^{-1} in $LiNbO_3$, from /108/.

scattering of the E(TO) phonon at 582 cm^{-1} due to reflections was much stronger than the polariton scattering efficiency in the forward direction. Photoelectric observation of the branch was impossible. The results obtained with the photographic method, however, were in good agreement with calculations by Obukhovskii et al. /110/. The main advantage of the method is obviously that scattered light can easily be integrated over several hours. Fig.35 shows the experimental data points derived from photometric measurements of the photographic plate. A feature of special interest observed in this figure is a Fermi resonance between the polariton and a second-order phonon at ∿ 537 cm^{-1}. The maximum frequency splitting recorded was 7 cm^{-1}, see also 5.5.

As illustrated in Fig.28b, there are scattering geometries in uniaxial and biaxial crystals that also allow the observation of the photon-like branches. The frequencies recorded again correspond to the intersection points of the dispersion curves and the hyperbolas determined by (4-57). For suitable refractive indices n_i and n_s there may even be two intersection points with the upper branch,

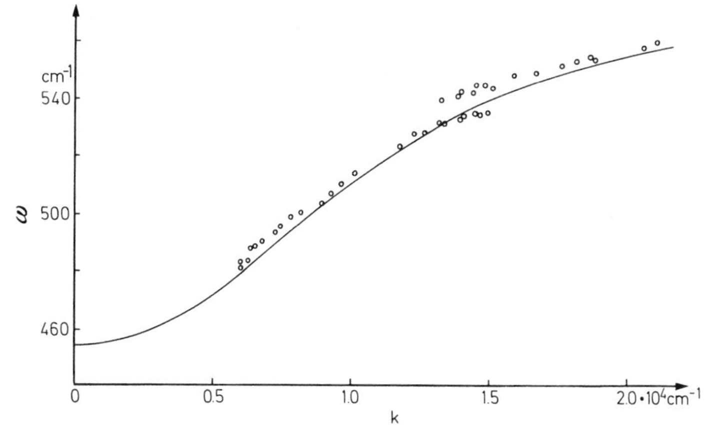

Fig.35 Calculated dispersion and experimental data corresponding to the ordinary polariton branch associated with the E(TO) phonon at 582 cm^{-1} in $LiNbO_3$. Fermi resonance takes place with a second-order phonon at ∿ 537 cm^{-1}. The maximum frequency splitting is about 7 cm^{-1}, see 5.5, from /108/.

as indicated in the figures. This happens in $LiNbO_3$ with, for instance, argon-laser (λ_i = 488 nm and λ_i = 514.5 nm) or He-Cd-laser excitation (λ_i = 441.6 nm). For $\phi = 0$ the limiting line determines the pair of the highest and lowest frequencies observed. Increasing scattering angles ϕ cause the lower polariton frequency to rise and the higher one to fall until at a certain angle $\phi = \phi_o$ the two coincide. For $\phi > \phi_o$, finally, the observation of the upper branch is impossible. The possibilities of such experiments have been discussed in detail by Obukhovskii et al. /110/.

Within the last few years light scattering owing to the higher-frequency intersection point has been reported by several authors on different materials /109/ and /111 through 117/. The phenomenon has become known as parametric luminescence. Klyshko and coworkers in 1970 /118/ were the first to show the connection with polariton scattering. They succeded in recording scattering processes corresponding to both intersection points simultaneously in $LiNbO_3$, using the photographic method described above. We show the result in a rerecorded spectrogram excited by a He-Cd laser, Fig.36. In this figure the ω axis is perpendicular to the exciting spectral line at the bottom. It can easily be seen that the polariton frequency of

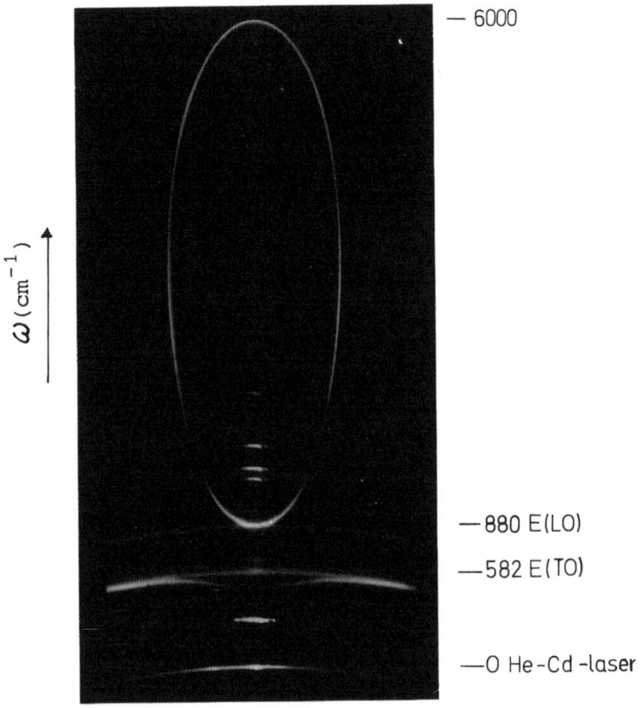

Fig.36

Spectrogram showing Raman scattering by the ordinary photon-like polariton branch, and parametric luminescence in $LiNbO_3$ recorded by Winter /119/.

the photon-like branch for straight forward scattering in the lower part almost coincides with the highest-frequency E(LO) phonon at 880 cm^{-1}. A second scattering process at $\sim$ 5000 cm^{-1} is observed simultaneously, in agreement with the prediction by Fig.31. For greater scattering angles ϕ the frequencies behave as described above. The smallest diameter of the oval figure corresponds to an external divergence of the scattered rays of $2\phi_o'$, see Figs.28b and 33. A quantitative comparison of experimental data with the calculated ordinary branch has been presented by Claus et al. /120/, Fig.37. Corresponding studies on $LiIO_3$ have been published by Winter /121/.

When the spectrogram stands vertically with the ω axis as shown in Fig.36, it can be regarded as a direct picture of the polariton dispersion curves. For this purpose all branches in Fig.31 have to be added symmetrically for negative $\underline{k}$ directions on the left side of the ω axis. The diagram then has to be cut along the limiting lines and folded together vertically in such a way that the upper two limiting lines and the lower ones coincide along the ω axis, which then corresponds to the symmetry axis of Fig.36.

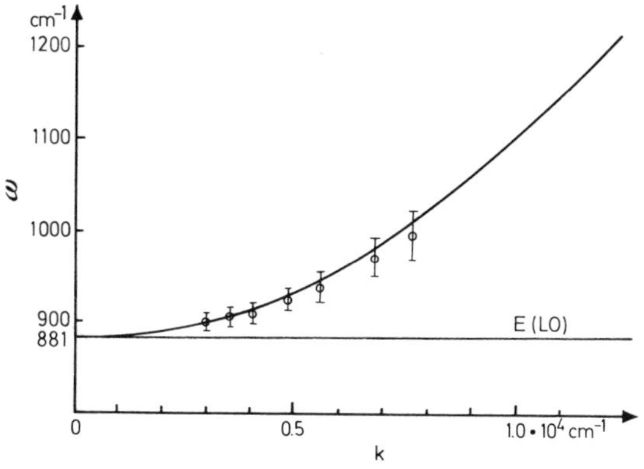

Fig.37 Calculated dispersion branch and experimental data concerning polaritons associated with the ordinary photons in $LiNbO_3$, from /120/.

Owing to light scattering as in Fig.28b, only the ordinary photon-like polariton branch seems to be observable. When oblique orientations are used, however, the extraordinary photon-like branches may also be observed. For decreasing angles θ between the wave-vector and the optic axis, they move toward the directionally independent ordinary branch. As a result intersection points may also occur between the limiting lines and these branches. Furthermore, variation of the direction of the extraordinary ray allows continuous variation of the corresponding refractive index from n_o to $n_{e(max)}$ in optically positive crystals and from $n_{e(min)}$ to n_o in optically negative media. Accordingly, the slopes of the limiting lines and the intersections with the ω axis can be changed, see (4-58), (4-59), and (4-60). Experiments of this kind have recently been carried out by Ponath and coworkers on $LiIO_3$ /122, 123, 353/ and by Winter on $LiNbO_3$ /124/.

We have seen from (4-53), (4-55) and (4-61) that extraordinary and ordinary photons as well as phonon-like polaritons degenerate for $\underline{k} \parallel z$, whereas for increasing angles θ the degeneracy is removed continuously. A simultaneous observation of ordinary and extraordinary polaritons associated with the same directional dispersion branch should therefore be possible for $\theta \neq 0$. Such experimental results were first reported by Claus and Winter /102/ relativ to the lowest-frequency directional dispersion branch in $LiNbO_3$, Figs.17 and 27c. For better orientation the position of the ordinary E(TO) phonon at 152 cm^{-1} is indicated by a dashed line in Fig.27c. For scattering angles $\phi > 3.6^{\circ}$, the ordinary polariton appears at $\sim$ 152 cm^{-1} and the extraordinary one at $\sim$ 198 cm^{-1}, which is the frequency of the extraordinary phonon for large k and $\theta \approx 82^{\circ}$, see Fig.17. Decreasing ϕ causes both polariton branches to move towards $\omega = 0$. Because of damping and broadening of the Raman lines in the region of strong dispersion caused by the finite aperture $\phi \pm (1/2)\Delta\phi$, ($\Delta\phi \approx 0.2^{\circ}$), the components can no longer be resolved for $\phi < 1.8^{\circ}$. Fig.38 shows the dispersion branches of the lower-frequency region in the same way they are recorded in the spectra Fig.27c. The ordinary E(TO) branch in question has been sketched as a dashed curve. This curve is identical to the lowest-frequency one in Fig.31. The ordinary E(TO) polaritons in Fig.27c are thus also identical to those reproduced in Fig.32. It should be mentioned that owing to some backward reflections inside the sample the 'quasi'-A_1(TO) phonon at 253 cm^{-1} again appears as a weak Raman line.

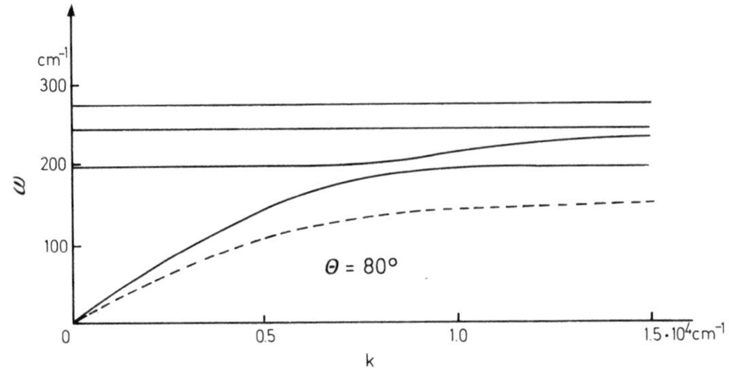

Fig.38 Dispersion branches of polaritons in $LiNbO_3$ in the lower-frequency region as they are observed in Fig.27c. The lowest-frequency ordinary branch is indicated by a dashed curve, from /102, 103/.

Additional Literature

Ohama, N., Okamoto, Y.: Polariton dispersion relation in cubic $BaTiO_3$ /218/.
Asawa, C.K., Barnoski, M.K.: Scattering from E_1 polaritons in $LiIO_3$ /224/.
Mavrin, B.N., Abramovich, T.E., Sterin, Kh.E.: Transverse polaritons in a $LiNbO_3$ crystal /233/.

4.9 POLARITONS IN BIAXIAL CRYSTALS

According to (3-20) and (4-21) the general dispersion relation of polaritons in orthorhombic crystals can be written

$$\varepsilon_1(\omega)(\omega^2\varepsilon_2(\omega)-c^2k^2)(\omega^2\varepsilon_3(\omega)-c^2k^2)s_1^2 +$$

$$\varepsilon_2(\omega)(\omega^2\varepsilon_3(\omega)-c^2k^2)(\omega^2\varepsilon_1(\omega)-c^2k^2)s_2^2 +$$

$$\varepsilon_3(\omega)(\omega^2\varepsilon_1(\omega)-c^2k^2)(\omega^2\varepsilon_2(\omega)-c^2k^2)s_3^2 = 0 \quad . \qquad (4\text{-}63)$$

For $k \to 0$, i.e. wave-vectors that are small compared with the polariton region ($k \lesssim 10^3\ cm^{-1}$), (4-63) reduces to

$$\omega^4\varepsilon_1\varepsilon_2\varepsilon_3\ s_1^2 + \omega^4\varepsilon_1\varepsilon_2\varepsilon_3\ s_2^2 + \omega^4\varepsilon_1\varepsilon_2\varepsilon_3\ s_3^2 = 0$$

and, because $s_1^2 + s_2^2 + s_3^2 = 1$,

$$\omega^4\varepsilon_1(\omega)\varepsilon_2(\omega)\varepsilon_3(\omega) = 0 \quad . \qquad (4\text{-}64)$$

The three dielectric functions may be replaced by the corresponding Kurosawa relations, (4-14). If $n_1 + n_2 + n_3 = n$ as before denotes the total number of polar modes, (4-64) is obviously of power $(n + 2)$ in ω^2. The corresponding solutions are: the double root, $\omega_{1,2}^2 = 0$, and the zeros of the three dielectric functions lead to n_1, n_2 and n_3 frequencies, which are identical to those of the exactly longitudinal phonons for the three principal directions. All modes except for the two lowest-frequency branches are without dispersion in the center of the 1BZ. This result is in agreement with the results for uniaxial crystals, see 4.7. The double root $\omega_{1,2} = 0$ in general corresponds to two different direction-dependent polariton branches going to zero energies for $k \to 0$. In uniaxial crystals one of these becomes the lowest-frequency, directionally independent ordinary dispersion branch and the other the corresponding extraordinary dispersion branch. Eq.(4-52) describing extraordinary polaritons is therefore only of power $(n_\perp + n_\parallel + 1)$ in ω^2. We also recall that the m + 1 polariton dispersion branches in cubic crystals described by (4-23) were all directionally independent and derived as doubly degenerate, a fact which there was without physical significance, see 4.3.

The counting of modes in cubic, uniaxial,and orthorhombic crystals may easily cause some confusion. We therefore briefly summarize the nomenclature used. In orthorhombic systems the total number of modes is $n_1 + n_2 + n_3 = n$, the integers 1, 2 and 3 referring to the three principal axes. According to (4-21), (4-63) and (4-64), the n phonon modes produce (n + 2) polariton branches. The two additional modes appear because for every wave-vector direction there are two photon-like branches. For $k \to \infty$ these describe pure photons and are thus not counted as phonons. In cubic crystals, $n_1 = n_2 = n_3 \equiv m$, see (4-22) and related text. The number of modes with different frequencies are identical for the three principal directions. Strictly speaking, (4-23) determines m + 2(m + 1) solutions, m of them being longitudinal, and the (m + 1) transverse modes are doubly degenerate for all directions. In uniaxial crystals, $n_1 = n_2 \equiv n_\perp \neq n_3 \equiv n_\parallel$. Eq.(4-36) determines $(n_\perp + 1) + (n_\perp + n_\parallel + 1)$ polariton branches. There are $(n_\perp + 1)$ ordinary and $(n_\perp + n_\parallel + 1)$ extraordinary modes. Following a (not very beautiful) convention, we referred in 4.6 to the 'total' number of extraordinary phonons as $n_\perp + n_\parallel = n$, which is not consistent with $n = n_1 + n_2 + n_3$, as defined earlier. However, this causes hardly any errors because the n_1 and n_2 modes along the principal directions in the optically isotropic plane are indistinguishable with respect to their frequencies. Similarly, the total number of phonons in cubic crystals have frequently been referred to as $n_1 = n_2 = n_3 \equiv n$ in the literature. It all depends on whether degeneracy is taken into account or not.

For a certain principal direction, for instance $\alpha = 1$, $s_1 = 1$, $s_2 = 0$ and $s_3 = 0$. Eq.(4-63) thus becomes

$$\varepsilon_1(\omega)(\omega^2\varepsilon_2(\omega) - c^2k^2)(\omega^2\varepsilon_3(\omega) - c^2k^2) = 0 \quad . \tag{4-65}$$

In order to determine the slopes of the two lowest-frequency polariton branches in the origin, we consider the $\omega \to 0$ limit of (4-65)

$$\varepsilon_{10}\,(\omega^2\varepsilon_{20} - c^2k^2)(\omega^2\varepsilon_{30} - c^2k^2) = 0 \tag{4-66}$$

which can easily be derived by taking into account the Kurosawa relation (4-14). The two brackets in (4-66) directly determine the slopes in question to be $\omega/k = c/\sqrt{\varepsilon_{20}}$ and $\omega/k = c/\sqrt{\varepsilon_{30}}$, respectively. Corresponding results are obtained for the other two directions, $\alpha = 2$ and $\alpha = 3$. The polariton phase velocities of the lowest branches in the origin are thus $c/\sqrt{\varepsilon_{\alpha 0}}$.

For $k \to \infty$, on the other hand, we obtain a description of polar phonons in the region $10^5 \lesssim k \lesssim 10^7\ \text{cm}^{-1}$. For finite ω, the first terms in the brackets of (4-63) are small, so that the dispersion relation reduces to

$$\varepsilon_1(\omega)s_1^2 + \varepsilon_2(\omega)s_2^2 + \varepsilon_3(\omega)s_3^2 = 0 \quad . \tag{4-67}$$

Because the wave normal vector $\underline{s} = (s_1, s_2, s_3)$ still appears in (4-67), all polar phonons in orthorhombic crystals in general show directional dispersion and there is no remaining dependence on the wave-vector magnitude. Taking the Kurosawa relation into account, we can verify that for a certain direction, e.g. $\alpha = 1$, the solutions of (4-67) are:

$\omega = \omega_{1Li}$, $i = 1 \ldots n_1$ = longitudinal modes

$\omega = \omega_{2Tj}$, $j = 1 \ldots n_2$ and

$\omega = \omega_{3T\ell}$, $\ell = 1 \ldots n_3$ = transverse modes.

In addition to these branches with finite frequencies, (4-63) describes two photon-like modes with infinite frequencies for $k \to \infty$, see above. The Kurosawa relation trivially requires $\varepsilon_\alpha(\omega) = \varepsilon_{\alpha\infty}$ for $\omega \to \infty$ and (4-63) accordingly becomes

$$\varepsilon_{1\infty}(\omega^2\varepsilon_{2\infty} - c^2k^2)(\omega^2\varepsilon_{3\infty} - c^2k^2) = 0 \tag{4-68}$$

for the direction $\alpha = 1$. From the zeros of the brackets the slopes of the two photon-like branches are derived as $\omega/k = c/\sqrt{\varepsilon_{2\infty}}$ and $\omega/k = c/\sqrt{\varepsilon_{3\infty}}$, respectively. Corresponding results are again obtained for the other principal directions, $\alpha = 2$ and $\alpha = 3$.

Only in the orthorhombic crystal classes D_2 (= 222) and C_{2v} (= mm2) are there polar modes which are simultaneously infrared- and Raman-active; materials belonging to the crystal class D_{2h} (= mmm) have an inversion center so that polaritons cannot be directly observed by light scattering. Unfortunately this happens also in D_2 type crystals where the form of Raman tensors prevents a direct observation, see Appdx 4.

To improve understanding, we discuss the phonon directional dispersion effects for a crystal with D_2 (= 222) symmetry in somewhat more detail. Directional dispersion is observable, see Fig.39a, b, c.

Polar B_1 modes have dipole moments in the direction $\alpha = 3$, whereas the B_2 and B_3 modes have their dipole moments lying in the directions $\alpha = 2$ and $\alpha = 1$, see Appdx 4. Exactly transverse B_1 phonons may therefore be observed only for wave-vector directions in a plane perpendicular to the direction $\alpha = 3$. Eq.(4-67) in this case becomes

$$\varepsilon_1(\omega) s_1^2 + \varepsilon_2(\omega) s_2^2 = 0 \qquad (4\text{-}69)$$

because $s_3 = 0$. The equation determines the frequencies of the remaining direction-dependent phonon modes. These modes are of mixed type ($B_2 + B_3$) for general wave-vector directions in the 1, 2 plane since only the two dielectric functions $\varepsilon_2(\omega)$ and $\varepsilon_3(\omega)$ are involved. Modes of pure type B_1(LO), B_2(LO), or B_3(LO), on the other hand, appear only for the directions $\alpha = 3$, $\alpha = 2$ and $\alpha = 1$, respectively. Like the B_1(TO) modes, which did not show any directional dispersion for wave-vectors lying in the 1, 2 plane, the B_2(TO) modes are not direction-dependent in the 1, 3 plane and the B_3(TO) modes in the 2, 3 plane. Mode-mixing with directional dispersion, however, takes place in each case for the other two types of phonons. These features are of great importance for the observation of polaritons. Thus, scattering experiments with wave-vector triangles lying in only one of the principal planes always allow the recording of polaritons associated with purely transverse phonons of one symmetry type without directional dispersion.

Experimental data concerning the directional dispersion of phonon modes originating from the vibrations of the SO_4 ions in orthorhombic $MgSO_4 \cdot 7H_2O$ have been carried out by Graf et al. /125/, for all three symmetry planes, see Fig.39a, b, c. The material belongs to the space group D_2^4 (= $P2_12_12_1$) discussed above. Similar, though less complete data have also been published for $NaNO_2$ /93/, benzophenone /126/, SbSI /127/, and α-HIO_3 /128 to 132/. Systematic measurements in the real polariton region $10^3 \lesssim k \lesssim 2 \times 10^4$ cm^{-1} have hitherto been published only for the purely transverse polar A_1, B_1, and B_2 modes in $KNbO_3$ by Winter, Claus et al. /137,317,350/. A spectra series showing the A_1(TO) polaritons in this material is reproduced in Fig.40a. Four dispersion branches can be observed: for decreasing internal scattering angles the Raman lines are shifted towards lower wave numbers each starting at a TO-phonon frequency and ending at the frequency of the next LO-phonon at lower wave numbers. Note that the lowest dispersion branch again moves towards $\omega = 0$ for $k \to 0$. Comparison of the experimental data with calculated dispersion curves shows good agreement, Fig.40b /317,350/. Similar but less complete data have also become known for α-HIO_3 /131, 132/ and $Ba_2NaNb_5O_{15}$ (banana) /133/.

In monoclinic systems the two crystal classes C_2 (= 2) and C_s (= m) show a lack of an inversion center. The only symmetry elements in these crystal classes are a twofold axis and a symmetry plane, respectively. According to convention /45/, these symmetry elements determine the y axis.

In materials with C_2 (= 2) symmetry, for instance, the totally symmetric vibrations all have their dipole moments in y direction whereas the polar modes of B type have dipole moments lying in different directions in the xz plane. Exactly longitudinal A(LO) modes are thus expected for wave-vectors propagating along the y axis and A(TO) phonons without directional dispersion for wave-vectors in the xz plane. For arbitrary angles θ between the wave-vector and the y axis, there are modes of mixed type (A + B) showing directional dispersion. When $\underline{k} \parallel y$, all modes of B type are exactly transverse. B(LO) modes, on the other hand, appear only for certain $\underline{k}$ directions in the xz plane determined by the dipole moments in question. B(TO) modes in the xz plane are observed one by one for wave-vector directions perpendicular to those of the B(LO) modes. When the wave-vector $\underline{k}$ rotates in the xz plane, the nontotally symmetric phonons do show directional dispersion because of coup-

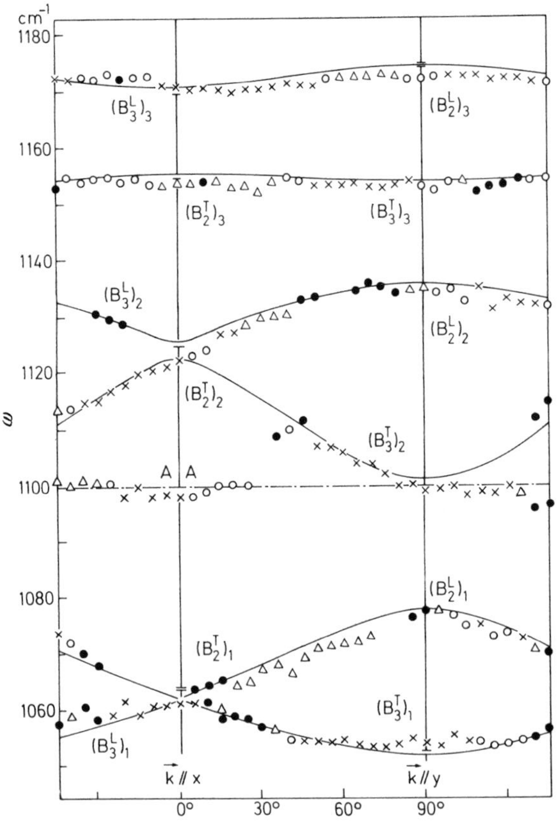

Fig.39a Directional dispersion of modes originating from the vibrations of SO_4 ions in orthorhombic Mg $SO_4 \cdot 7\ H_2O$. The principal directions α = 1,2, and 3 are indicated by x, y, and z, from /125/.

lings between the different modes of B type.

Every single B vibration determines a plane containing its dipolemoment vector and the y axis where it couples with an A mode. All other phonons with arbitrary wave-vector directions in this plane are directionally dependent in general because of multimode mixing involving an A phonon and other B modes. At the same time every B mode determines a plane perpendicular to its dipolemoment. The corresponding exactly transverse B(TO) phonon is observed for all wave-vectors without directional dispersion in this plane.

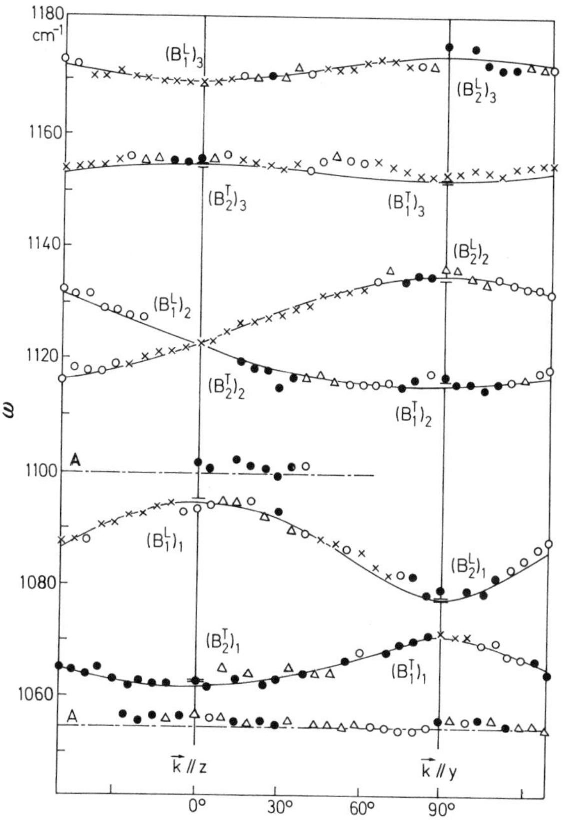

Fig.39b Directional dispersion of modes originating from the vibrations of SO_4 ions in orthorhombic $MgSO_4 \cdot 7H_2O$. The principal directions α = 1,2, and 3 are indicated by x, y, and z, from /125/.

Multimode mixing seems to be the most significant feature of biaxial crystals in the same way as extraordinary modes (two-mode couplings) were characteristic of uniaxial crystals. Note that multimode mixing also takes place in orthorhombic materials for planes other than the three symmetry planes.

Systematic experiments in monoclinic crystals may be performed as follows. The frequencies of B(TO) phonons can be recorded for wavevector directions $\underline{k} \parallel y$. The fixed $\underline{k}$ directions in the xz plane for different B(TO) modes can then be determined from directional dispersion measurements because their frequencies are known. The corresponding longitudinal modes propagate perpendicular to the directions determined in this way. Consequently, we know which planes

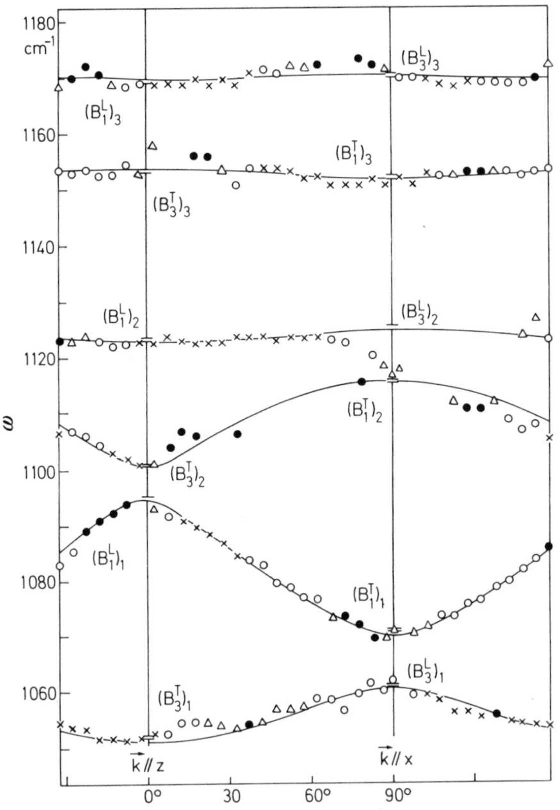

Fig.39c Directional dispersion of modes originating from the vibrations of SO_4 ions in orthorhombic Mg $SO_4 \cdot 7\ H_2O$. The principal directions α = 1,2, and 3 are indicated by x, y, and z, from /125/.

allow directional dispersion-free investigations of polariton dispersion of purely transverse B(TO) branches. Investigations of this kind, however, have not yet been undertaken to the authors' knowledge.

In triclinic systems there is no symmetry element allowing a classification of different vibrational modes. The polar phonons in such materials have their dipole moments lying in different directions in space. Every mode again determines a plane perpendicular to its dipole moment. The frequency of the exactly transverse phonon for all directions appears in this plane. The corresponding longitudinal phonon, on the other hand, appears only for wave-vectors propagating parallel to the dipole moment. All other modes in general

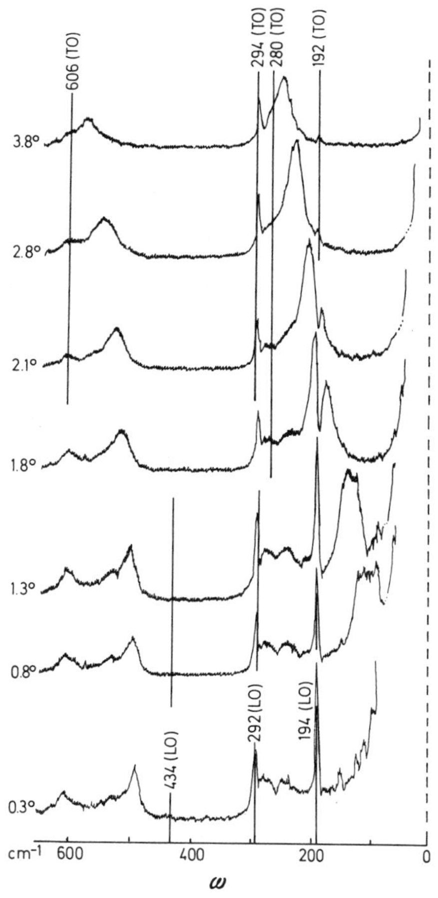

Fig.40a Spectra series of A_1(TO) polaritons in orthorhombic $KNbO_3$. Internal scattering angles are given to the left of each spectrum. The scattering geometry y(z z)y corresponds to $\phi=0^o$. Scattering plane: xy, from /137/.

show directional dispersion due to multimode couplings in the plane cited. The difference from monoclinic crystals is that this statement now holds for all waves and not only for a certain group such as B modes in C_2 (= 2) crystals. Furthermore, all dipole moments are lying in arbitrary directions in space and not only in one plane as in monoclinic crystals. No detailed experimental or theoretical studies have yet been made in this field.

Additional Literature

Asawa, C.K.: Frequency versus wave-vector for a diatomic ionic orthorhombic biaxial crystal /223/.

Graf, L., Schaack, G., Unger, B.: Raman scattering of polaritons in uniaxial and biaxial piezoelectric crystals /236/.

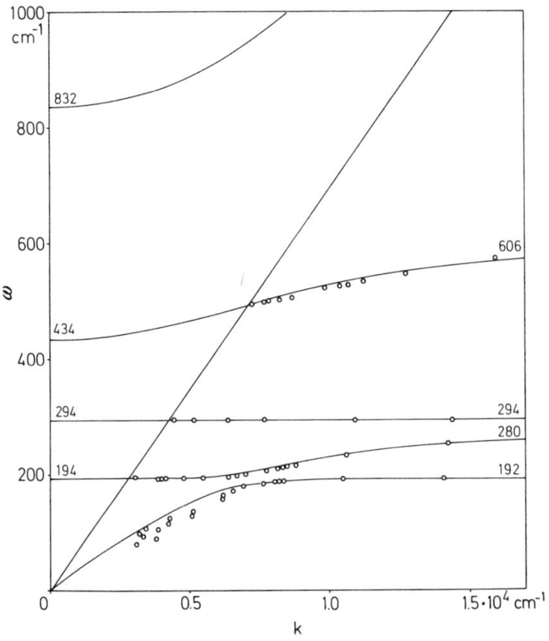

Fig.40b Experimental data and calculated dispersion curves of A_1(TO) polaritons in $KNbO_3$, as shown in Fig.40a, from /317/.

Krauzman, M., Postollec, M. le, Mathieu, J.P.: Vibration spectra, structure and angular dispersion of phonons in crystalline iodic acid (α-HIO_3) /237/.

Belikova, G.S., Kulevsky, L.A., Polivanov, Yu.N., Poluektov, S.N., Prokhorov, K.A., Shigoryn, V.D., Shipulo, G.P.: Light scattering from polaritons in m-dinitrobenzene single crystal /346/.

4.10 DAMPING OF POLARITONS

In the foregoing sections polaritons have been considered in the harmonic approximation, i.e. in the lattice potential energy (4-4), only quadratic terms have been taken into account, terms of higher order being neglected. Furthermore, all perturbations such as point defects, dislocations, and surfaces have been left out of consideration. These perturbations of the harmonic lattice potential limit the lifetime of polaritons and cause damping. For many features, such as the half-widths of Raman lines, the losses in the stimulated Raman effect, and the structures of the infrared spectra, damping play an important role.

A rigorous theory of damping should include all the different kinds of perturbations quantitatively, i.e. the origin of damping should be expressed by means of the microscopic mechanism. Appropriate theories, in so far as they exist, contain so many unknown parameters that quantitative calculations and comparison with experimental data are still impossible except for the most trivial materials.

Furthermore, we have to consider that any damping mechanism affecting one of the basic quantities in the fundamental equations (4-7), (4-8) and (4-9) will cause damping of the entire polariton system. A damping mechanism thus can act via a quasinormal coordinate $Q_{\alpha j}$, via the electric field $\underline{E}$, via the polarization $\underline{P}$, or via any of these parameters simultaneously.

As has been shown by various authors, damping mechanisms originate from anharmonic interactions /140, 141/ as well as from interactions with impurities /142/. A frequently considered ansatz introduces the damping factor as frequency-dependent, see /143-145, 148, 149/. We also refer the reader to papers by Maradudin and coworkers /293-296/, Cowley /297/, and Benson and Mills /150/.

The microscopic models developed hitherto do not yet seem to have reached a final form where experimental studies could verify them. We therefore present a phenomenologic theory for damped polaritons which does not depend on the microscopic mechanism acting in any special case.

In common crystalline materials a certain direction and its opposite are equivalent. We therefore assume that both $\omega_i(\underline{k})$ and $-\omega_i^*(\underline{k})$ represent polariton frequencies, i.e. the waves can always be combined into pairs $(\omega_i, -\omega_i^*)$.

By analogy with the Kurosawa relation for polaritons with real frequencies (4-14), a generalized corresponding relation can be derived for damped waves. In this case

$$\varepsilon_\alpha(\omega) = \varepsilon_{\alpha\infty} \prod_{j=1}^{n_\alpha} \frac{(\omega_{\alpha Lj}-\omega)(-\omega^*_{\alpha Lj}-\omega)}{(\omega_{\alpha Tj}-\omega)(-\omega^*_{\alpha Tj}-\omega)} \quad , \tag{4-70}$$

where $\omega_{\alpha Lj} = \bar{\omega}_{\alpha Lj}-\gamma_{\alpha Lj}/2$ and $\omega_{\alpha Tj} = \bar{\omega}_{\alpha Tj}-\gamma_{\alpha Tj}/2$. $\bar{\omega}_{\alpha Lj}$ and $\bar{\omega}_{\alpha Tj}$ denote the (real) frequencies, and $\gamma_{\alpha Lj}$ and $\gamma_{\alpha Tj}$ the damping factors of the purely longitudinal and transverse optic phonons polarized in α direction. The real frequencies correspond as before to undamped polaritons in the limiting case $k \to \infty$. Eq.(4-72) can be written in the equivalent form

$$\varepsilon_\alpha(\omega) = \varepsilon_{\alpha\infty} \prod_{j=1}^{n_\alpha} \frac{(|\omega_{\alpha Lj}|^2-i\omega\gamma_{\alpha Lj}-\omega^2)}{(|\omega_{\alpha Tj}|^2-i\omega\gamma_{\alpha Tj}-\omega^2)} \quad . \tag{4-71}$$

A similar result was derived by Barker in 1964 /63/ for the principal directions of a uniaxial polyatomic crystal. The polariton dispersion relation is derived in the form of a generalized Fresnel equation as for undamped polaritons, see 4.2:

$$\varepsilon_1(\varepsilon_2-n^2)(\varepsilon_3-n^2)s_1^2+\varepsilon_2(\varepsilon_3-n^2)(\varepsilon_1-n^2)s_2^2+\varepsilon_3(\varepsilon_1-n^2)(\varepsilon_2-n^2)s_3^2 = 0 \quad . \tag{4-72}$$

The dielectric functions $\varepsilon_1(\omega)$, $\varepsilon_2(\omega)$, and $\varepsilon_3(\omega)$, however, are determined by either (4-70) or (4-71).

In uniaxial crystals Fresnel's equation again splits into two parts, the first part describing ordinary damped polaritons

$$(c^2k^2/\omega^2) = n^2 = \varepsilon_\perp \tag{4-73}$$

and the second the extraordinary modes

$$(c^2k^2/\omega^2) = n^2 = \frac{\varepsilon_\perp\varepsilon_\parallel}{\varepsilon_\perp s_\perp^2 + \varepsilon_\parallel s_\parallel^2} \quad . \tag{4-74}$$

The indices and abbreviations are the same as those introduced in 4.4. All 'damped' dispersion branches $\omega_i(\underline{k})$ for arbitrary wave-vectors $\underline{k}$ can be calculated provided the numerical values of the $\bar{\omega}_{\alpha Lj}$, $\bar{\omega}_{\alpha Tj}$, $\gamma_{\alpha Lj}$, $\gamma_{\alpha Tj}$, and $\varepsilon_{\alpha\infty}$ are known, see /146/.

We distinguish between spatial, temporal, and mixed spatial-temporal damping. For pure spatial damping the wave-vector $\underline{k}$ is supposed to be complex

$$\underline{k} = \mathrm{Re}(\underline{k}) + i\mathrm{Im}(\underline{k}) \tag{4-75}$$

whereas the frequency ω_i is real. On the contrary, the frequency is made complex for pure temporal damping

$$\omega_i(\underline{k}) = \bar{\omega}_i(\underline{k}) + i\gamma_i(\underline{k})/2 \quad , \tag{4-76}$$

where the $\bar{\omega}_i(\underline{k}) = \mathrm{Re}\omega_i(\underline{k})$ denote the frequencies and the $\gamma_i(\underline{k}) = 2\,\mathrm{Im}\omega_i(\underline{k})$ the temporal damping factors of the waves. The magnitudes of the complex frequencies ω_i are $|\omega_i| = (\omega_i^2 + \gamma_i^2/4)^{1/2}$.

Numerical calculations have been carried out for ZnF_2 by Merten and Borstel /139/. Fig.41 shows the dispersion curves of spatially damped extraordinary polaritons for different wave-vector directions in this material. θ denotes in the usual way the angle between the optic axis and $\underline{k}$. The values on the abscissa are directly related to the refractive index n by

$$\mathrm{Re}\, n = (c/\omega)\mathrm{Re}\, k \quad \text{and} \quad \mathrm{Im}\, n = (c/\omega)\mathrm{Im}\, k \quad . \tag{4-77}$$

Pure spatial damping can be experimentally observed by IR absorption whereas the Raman effect essentially corresponds to temporal damping.

It can be shown /146, 147/ that for the damping factors of the polariton branches the sum rule holds:

$$\sum_{i=1}^{n+2} \gamma_i(\underline{k}) = \sum_{j=1}^{n_1} \gamma_{1Lj} + \sum_{k=1}^{n_2} \gamma_{2Lk} + \sum_{\ell=1}^{n_3} \gamma_{3L\ell}$$

$$= \sum_{j=1}^{n_1} \gamma_{1Tj} + \sum_{k=1}^{n_2} \gamma_{2Tk} + \sum_{\ell=1}^{n_3} \gamma_{3T\ell} = \text{const.} \tag{4-78}$$

n_α stands for the number of polar phonons in the principal direction α, $\alpha = 1, 2, 3$, and $n = n_1 + n_2 + n_3$ again denotes the total number of infrared-active optical phonons. The sum of the damping factors for all polariton branches is constant and independent of $\underline{k}$.

Specializing (4-78) for uniaxial crystals, we obtain a sum rule for ordinary polaritons

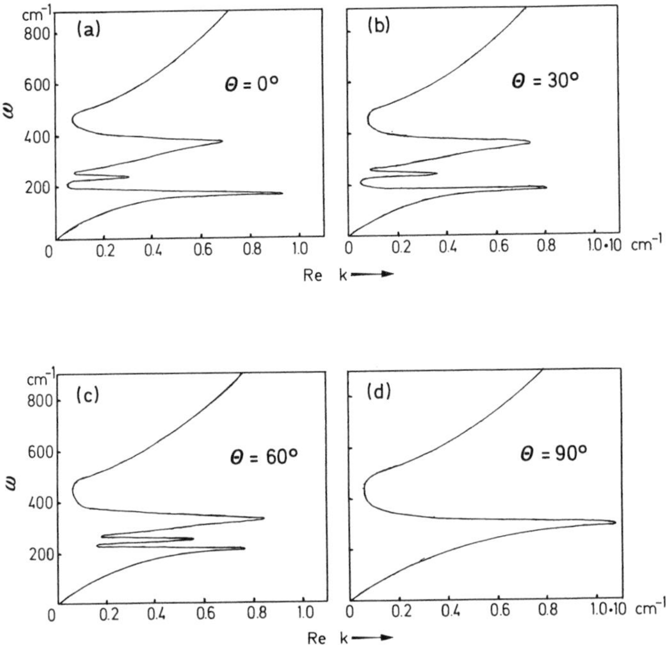

Fig.41 Dispersion curves $\omega=\omega(\mathrm{Re}\ k)$ for different angles θ between the wave-vector and the optic axis in uniaxial ZnF_2, from /139/.

$$\sum_{i=1}^{n_\perp+1} \gamma_{o,i}(k) = \sum_{j=1}^{n_\perp} \gamma_{\perp Lj} = \sum_{j=1}^{n_\perp} \gamma_{\perp Tj} = \mathrm{const.} \tag{4-79}$$

and another for the extraordinary modes

$$\sum_{i=1}^{n_\parallel+n_\perp+1} \gamma_{eo,i}(\underline{k}) = \sum_{j=1}^{n_\perp} \gamma_{\perp Lj} + \sum_{k=1}^{n_\parallel} \gamma_{\parallel Lk} = \sum_{j=1}^{n_\perp} \gamma_{\perp Tj} + \sum_{k=1}^{n_\parallel} \gamma_{\parallel Tk} = \mathrm{const.} \tag{4-80}$$

For the doubly degenerate transverse polaritons in cubic crystals we have

$$\sum_{i=1}^{m+1} \gamma_{Ti}(k) = \sum_{j=1}^{m} \gamma_{Lj} = \sum_{j=1}^{m} \gamma_{Tj} = \mathrm{const.} \tag{4-81}$$

in analogy to the ordinary polaritons in uniaxial crystals. m denotes the number of infrared-active phonons (single count). It (trivially) follows that the damping of all purely longitudinal branches is $\underline{k}$-independent. The sum rules have not yet been experimentally verified.

The theory outlined above does not depend in detail on the microscopic mechanism or any special form of the damping terms. If damping is assumed to be proportional to the normal coordinates, the generalized first fundamental polariton equation, (4-5), becomes /63, 159/

$$\ddot{\underline{Q}} + \Gamma(\omega) \cdot \dot{\underline{Q}} = B^{11} \cdot \underline{Q} + B^{12} \cdot \underline{E} \quad , \tag{4-82}$$

where $\Gamma(\omega)$ denotes a damping tensor. With a plane wave ansatz, the total set of fundamental equations for damped polaritons becomes

$$-\omega^2 \underline{Q} = (B^{11} + i\omega\Gamma(\omega)) \cdot \underline{Q} + B^{12} \cdot \underline{E} \quad , \tag{4-83}$$

$$\underline{P} = (B^{12})^{+} \cdot \underline{Q} + B^{22} \cdot \underline{E} \quad , \tag{4-84}$$

$$\underline{E} = 4\pi(n^2-1)^{-1}\left[\underline{P} - n^2 \underline{s}(\underline{s} \cdot \underline{P})\right] \tag{4-85}$$

in analogy to (4-7), (4-8) and (4-9). The dispersion relation is derived in the same way as in 4.2 in the form of a generalized Fresnel equation, (4-72). If the diagonal tensor $\Gamma(\omega)$ is approximately constant in the vicinity of the resonance frequencies, the diagonal elements $\Gamma_{\alpha j}$ are identical with the $\gamma_{\alpha Tj}$ introduced above. This happens because the equations of motion of purely transverse modes for different normal coordinates are decoupled in the limiting case $\underline{k} \to \infty$. (4-83) then becomes explicitly

$$-\omega^2 Q_{\alpha j} = B^{11}_{\alpha j} Q_{\alpha j} + i\omega\Gamma_{\alpha j} Q_{\alpha j} \tag{4-86}$$

or, when $B^{11}_{\alpha j} = -\omega^2_{\alpha tj}$ is taken into account, see 4.2,

$$\omega^2 + i\omega\Gamma_{\alpha j} - \omega^2_{\alpha tj} = 0 \quad . \tag{4-87}$$

The $\omega_{\alpha tj}$ denote the frequencies of the undamped transverse polaritons in the limit $\underline{k} \to \infty$. We then easily derive the solutions

$$\omega_{\alpha Tj} = (i\Gamma_{\alpha j}/2) \pm (\omega^2_{\alpha tj} - \Gamma^2_{\alpha j}/4)^{1/2} \tag{4-88}$$

where

$$\bar{\omega}_{\alpha Tj} = \mathrm{Re}(\omega_{\alpha Tj}) = \pm(\omega^2_{\alpha tj} - \Gamma^2_{\alpha j}/4)^{1/2} \qquad (4\text{-}89)$$

and

$$\gamma_{\alpha tj} = 2\,\mathrm{Im}(\omega_{\alpha tj}) = \Gamma_{\alpha j} \quad . \qquad (4\text{-}90)$$

These relations, however, hold only when the damping term has the form introduced in (4-82). The $\gamma_{\alpha tj}$ are there defined as independent of any special damping mechanism.

To date, theories leading to an explicit damping function from a special microscopic mechanism have been rare. As an example, we cite a relation derived by Maradudin and Wallis /296/. The authors have considered anharmonic couplings in a cubic crystal with only one TO phonon

$$\varepsilon(\omega) = \varepsilon_o + 4\pi\rho(\omega_T^2 - \omega^2 - 2\omega_T \Pi_R(\omega) - i2\omega_T \Pi_I(\omega))^{-1} \quad . \qquad (4\text{-}91)$$

$\Pi_R(\omega)$ and $\Pi_I(\omega)$ denote the real and imaginary parts of the phonon's self-energy for $\underline{k} = 0$.

Experiments concerning polariton damping are also rare. The damping function below $\omega = \omega_T$, for instance, has been investigated in GaP /143, 145/.

The phenomenological theory outlined so far assumes that the polaritons exist in the crystal as either undamped or damped waves. Polaritons, however, may also be coupled to dipole relaxations. Couplings of this kind are of great importance in ferroelectric materials because of the existence of Debye relaxations.

Relaxations are distinguished from the oscillating states by the fact that their 'frequencies' are purely imaginary, $\omega_i(\underline{k}) = i\gamma_i(\underline{k})/2$. By introducing a relaxation time $\tau_i = 2/\gamma_i$, we obtain a time dependence of the form

$$\exp(i\omega_i t) = \exp(-t/\tau_i) \quad . \qquad (4\text{-}92)$$

The coupling with Debye relaxations causes strong shifts of the real frequencies and strong damping of polaritons and can again be completely described by a phenomenologic theory. In contrast to the theory outlined above, at least one frequency now has to be introduced as purely imaginary. For the simplest case with only one relaxation polarized in α direction, imaginary LO and TO frequencies are introduced in the Kurosawa relation as

$$\omega_{\alpha LO} = i\gamma_{\alpha LO}/2 = i/\tau_{\alpha LO} \tag{4-93}$$

and

$$\omega_{\alpha TO} = i\gamma_{\alpha TO}/2 = i/\tau_{\alpha TO} \quad . \tag{4-94}$$

We then obtain

$$\varepsilon_\alpha(\omega) = \varepsilon_{\alpha\infty} \frac{\omega_{\alpha LO}-\omega}{\omega_{\alpha TO}-\omega} \prod_{j=1}^{n_\alpha} \frac{(\omega_{\alpha Lj}-\omega)(-\omega^*_{\alpha Lj}-\omega)}{(\omega_{\alpha Tj}-\omega)(-\omega^*_{\alpha Tj}-\omega)} \tag{4-95}$$

or

$$\varepsilon_\alpha(\omega) = \varepsilon_{\alpha\infty} \frac{\tau_{\alpha TO}(1+i\omega\tau_{\alpha LO})}{\tau_{\alpha LO}(1+i\omega\tau_{\alpha TO})} \prod_{j=1}^{n_\alpha} \frac{(|\omega_{\alpha Lj}|^2-i\omega\gamma_{\alpha Lj}-\omega^2)}{(|\omega_{\alpha Tj}|^2-i\omega\gamma_{\alpha Tj}-\omega^2)} \tag{4-96}$$

or

$$\varepsilon_\alpha(\omega) = \varepsilon_{\alpha\infty} \frac{\tau_{\alpha TO}(1+i\omega_{\alpha LO})}{\tau_{\alpha LO}(1+i\omega_{\alpha TO})} \prod_{j=1}^{n_\alpha} \frac{(\omega_{\alpha Lj}-\omega)(-\omega^*_{\alpha Lj}-\omega)}{(\omega_{\alpha Tj}-\omega)(-\omega^*_{\alpha Tj}-\omega)} \quad . \tag{4-97}$$

If there is more than one relaxation, an additional factor $(\tau_{\alpha Th}/\tau_{\alpha Lh})(1+i\omega\tau_{\alpha Lh})/(1+i\omega\tau_{\alpha Th})$ appears. If we denote the total number of relaxations by m_α, the correspondingly generalized Kurosawa relation becomes

$$\varepsilon_\alpha(\omega) = \varepsilon_{\alpha\infty} \prod_{h=1}^{m_\alpha} \frac{\tau_{\alpha Th}(1+i\omega\tau_{\alpha Lh})}{\tau_{\alpha Lh}(1+i\omega\tau_{\alpha Th})} \prod_{j=1}^{n_\alpha} \frac{(|\omega_{\alpha Lj}|^2-i\omega\gamma_{\alpha Lj}-\omega^2)}{(|\omega_{\alpha Tj}|^2-i\omega\gamma_{\alpha Tj}-\omega^2)} \quad . \tag{4-98}$$

The generalized Lyddane-Sachs-Teller relation derived from (4-98) for $\omega \to 0$ is

$$\varepsilon_{\alpha o} = \varepsilon_{\alpha\infty} \prod_{h=1}^{m_\alpha} \frac{\tau_{\alpha Th}}{\tau_{\alpha Lh}} \prod_{j=1}^{n_\alpha} \frac{|\omega_{\alpha Lj}|^2}{|\omega_{\alpha Tj}|^2} \quad . \tag{4-99}$$

Similar relations have been derived by Chaves and Porto /157/. A decomposition of (4-98) into partial fractions finally gives

$$\varepsilon_\alpha(\omega) = \varepsilon_{\alpha\infty} + \sum_{h=1}^{m_\alpha} \frac{C_{\alpha h}}{1+i\omega\tau_{\alpha Th}} + \sum_{j=1}^{n_\alpha} \frac{4\pi\rho_{\alpha j}}{|\omega_{\alpha Tj}|^2 - i\omega\gamma_{\alpha Tj} - \omega^2} \quad . \tag{4-100}$$

All real parameters ($\varepsilon_{\alpha\infty}$, $C_{\alpha h}$, $\tau_{\alpha Th}$, $|\omega_{\alpha Tj}|$, $\rho_{\alpha j}$, and $\gamma_{\alpha Tj}$) in this relation can be experimentally determined and consequently the generalized Kurosawa relation (4-98) completely determines the dielectric function.

The dispersion relation of polaritons including relaxations is obtained when substituting the dielectric functions $\varepsilon_\alpha(\omega)$ given by (4-98) or (4-100) into Fresnel's equation (4-72).

A generalized LST relation of the type of (4-99) has been derived also by Petersson and Müser from the thermodynamics of irreversible processes /298, 299, 345/. These authors in addition have shown that the $\tau_{\alpha Lh}$ and $\tau_{\alpha Th}$ can be arranged as

$$0 < \tau_{\alpha L1} \leq \tau_{\alpha T1} \leq \tau_{\alpha L2} \leq \tau_{\alpha T2} \leq \dots \leq \tau_{\alpha Tm_\alpha} < \infty \quad . \tag{4-101}$$

(The nomenclature used in /298/ corresponds to ours as follows: $\tau_{\alpha Th} \equiv \tau_{\varepsilon\alpha}$, $\tau_{\alpha Lh} \equiv \tau_{\beta\alpha}$, $\varepsilon_o \equiv 1/\beta^o$ and $\varepsilon_\infty \equiv 1/\beta^\infty$.)

As a result, we have for the additional factor in (4-99)

$$\prod_{h=1}^{m_\alpha} (\tau_{\alpha Th}/\tau_{\alpha Lh}) > 1 \quad . \tag{4-102}$$

Consequently, $\varepsilon_{\alpha o}$ appearing in the LST relation (4-15) will be raised by Debye relaxations.

In uniaxial ferroelectric crystals the optic axis coincides with the ferroelectric axis. Applying the relations derived above to the simplest type of ferroelectric material thus provides dispersion relations for ordinary and extraordinary polaritons

$$(c^2k^2/\omega^2) = \varepsilon_\perp(\omega) \tag{4-103}$$

and

$$(c^2k^2/\omega^2) = \frac{\varepsilon_\perp \varepsilon_\parallel}{\varepsilon_\perp s_\perp^2 + \varepsilon_\parallel s_\parallel^2} \quad , \tag{4-104}$$

respectively, with

$$\varepsilon_\perp(\omega) = \varepsilon_{\perp\infty} \prod_{j=1}^{n_\perp} \frac{|\omega_{\perp Lj}|^2 - i\omega\gamma_{\perp Lj} - \omega^2}{|\omega_{\perp Tj}|^2 - i\omega\gamma_{\perp Tj} - \omega^2} \tag{4-105}$$

and

$$\varepsilon_\parallel(\omega) = \varepsilon_{\parallel\infty} \frac{\tau_{\parallel T}(1+i\omega\tau_{\parallel L})}{\tau_{\parallel L}(1+i\omega\tau_{\parallel T})} \prod_{j=1}^{n_\parallel} \frac{|\omega_{\parallel Lj}|^2 - i\omega\gamma_{\parallel Lj} - \omega^2}{|\omega_{\parallel Tj}|^2 - i\omega\gamma_{\parallel Tj} - \omega^2} \quad , \tag{4-106}$$

if only one Debye relaxation exists and

$$\varepsilon_\parallel(\omega) = \varepsilon_{\parallel\infty} \prod_{h=1}^{m_\parallel} \frac{\tau_{\parallel Th}(1+i\omega\tau_{\parallel Lh})}{\tau_{\parallel Lh}(1+i\omega\tau_{\parallel Th})} \prod_{j=1}^{n_\parallel} \frac{|\omega_{\parallel Lj}|^2 - i\omega\gamma_{\parallel Lj} - \omega^2}{|\omega_{\parallel Tj}|^2 - i\omega\gamma_{\parallel Tj} - \omega^2} \tag{4-107}$$

if the total number of relaxations is $m_\parallel$. In uniaxial crystals Debye relaxations thus influence only the extraordinary modes. As described by (4-104), we refer to them as 'extraordinary Debye relaxations'. The dispersion of extraordinary Debye relaxations has recently been discussed by Merten and coworkers /154/.

For model theories concerning the coupling of polaritons to Debye relaxations, we refer the reader to /151-153, 156/.

The coupling of Debye relaxations with phonons and polaritons plays an important role in the dynamics of phase transitions of ferroelectric crystals (soft modes).

Additional Literature

Puthoff, H.E., Pantell, R.H., Huth, B.G., Chacon, M.A.: Near-forward Raman scattering in $LiNbO_3$ /214/.

Ohtaka, K., Fujiwara, T.: Polaritons in anharmonic crystals /217/.

Giallorenzi, T.G.: Quantum theory of light scattering by damped polaritons /230/.

Reinisch, R., Paraire, N., Biraud-Laval, S.: Etude de la courbe de dispersion des polaritons excités par diffusion Raman en présence d'amortissement /231/.

Ushioda, S., McMullen, J.D., Delaney, M.J.: Damping mechanism of polaritons in GaP /239/.
Inoue, M.: Dielectric dispersion formula for ferroelectrics /243/.

4.11 POLARITON EIGENVECTORS

For many problems, we are interested, in addition to the frequencies, in the eigenvectors themselves, such as the quasinormal coordinates $Q_{\alpha j}$, the electric-field components E_α, and the components of the macroscopic polarization P_α. When one is calculating, for instance, the Raman scattering cross-sections on the basis of (4-124), (4-125) and (4-126) it is necessary to know the quasinormal coordinates $Q_{\alpha j}$ and the electric field components E_α.

In order to determine the eigenvectors from the fundamental set of equations (4-7) through (4-9), the well-known methods of linear algebra can in principle be applied. The numerical values of the required eigenfrequencies can be obtained from the generalized Fresnel equation (4-20). This method has been applied to $LiNbO_3$ by Merten and Borstel /70, 72/ and to $GeSO_4$ by Unger and Schaack /158/. The numerical calculations of Borstel and Merten have shown that the behavior of the eigenvectors in the surroundings of 'crossing points' is of special interest.

As could recently be shown, however, the eigenvectors may also be derived analytically /159/. The derivation is based on the fundamental equation of motion (4-10) written in components

$$-\omega_m^2(\underline{k})\,Q_{\alpha j}^{(m)} = B_{\alpha j}^{11}Q_{\alpha j}^{(m)} + B_{\alpha j}^{12}E_\alpha^{(m)} \quad . \tag{4-108}$$

Here the index m denotes a certain polariton branch and α the principal direction, as before. The eigenfrequency $\omega_m(\underline{k})$ of the m-th polariton mode is supposed to be known from Fresnel's equation (4-20). From (4-108) we therefore obtain

$$Q_{\alpha j}^{(m)}(\underline{k}) = \frac{B_{\alpha j}^{12}E_\alpha^{(m)}(\underline{k})}{-B_{\alpha j}^{11}-\omega_m^2(\underline{k})} \quad . \tag{4-109}$$

By substituting $B_{\alpha j}^{11} = -\omega_{\alpha Tj}^2$, $B_{\alpha j}^{12} = \omega_{\alpha Tj}\sqrt{\rho_{\alpha j}}$ and taking (4-16) into account, we can write

$$Q_{\alpha j}^{(m)}(\underline{k}) = \frac{\sqrt{\rho_{\alpha j}}\ \omega_{\alpha Tj} E_{\alpha}^{(m)}(\underline{k})}{\omega^2_{Tj} - \omega_m^2(\underline{k})} = \tag{4-110}$$

$$= (\varepsilon_{\alpha\infty}/4\pi)\ \frac{\prod_{i=1}^{n_\alpha} (\omega_{\alpha Li}^2 - \omega_{\alpha Ti}^2)}{\prod_{\substack{i=1\\ i\neq j}}^{n_\alpha} (\omega_{\alpha Ti}^2 - \omega_{\alpha Tj}^2)} \cdot \frac{E_{\alpha}^{(m)}(\underline{k})}{(\omega_{\alpha Tj}^2 - \omega_m^2(\underline{k}))} \quad . \tag{4-111}$$

Eq.(4-109) has been derived by Loudon /73/ and others in an equivalent form.

The components of the electric field are correlated by

$$E_{\alpha}^{(m)} = \left[s_{\alpha}(n^2-\varepsilon_{\beta})/s_{\beta}(n^2-\varepsilon_{\alpha})\right] E_{\beta}^{(m)} \quad . \tag{4-112}$$

This is a well-known result from crystal optics, see the relevant discussion in 3.2. Eq.(3-17) can be written $(\varepsilon(\omega)-n^2 I+n^2\underline{s}\ \underline{s})\cdot\underline{E} = 0$, where I denotes the unit matrix. If the equation is split into components, (4-112) can easily be derived by applying Cramer's rule /161/. s_α and s_β are the direction cosines and ε_α and ε_β the corresponding dielectric functions, (4-14)

$$\varepsilon_{\alpha}(\omega) = \varepsilon_{\alpha\infty} \prod_{j=1}^{n_\alpha} \frac{\omega_{\alpha Lj}^2 - \omega_m^2(\underline{k})}{\omega_{\alpha Tj}^2 - \omega_m^2(\underline{k})} \quad . \tag{4-113}$$

By combining (4-111) and (4-112) it is possible to express all quasinormal coordinates as a function of only one component of $\underline{E}^{(m)}$, which shall be $E_1^{(m)}(\underline{k})$. We then obtain

$$Q_{1j}^{(m)}(\underline{k}) = \frac{B_{1j}^{12} E_1^{(m)}(\underline{k})}{\omega_{1Tj}^2 - \omega_m^2(\underline{k})} \quad , \tag{4-114}$$

$$Q_{2j}^{(m)}(\underline{k}) = \frac{s_2(n^2-\varepsilon_1)}{s_1(n^2-\varepsilon_2)} \cdot \frac{B_{2j}^{12} E_1^{(m)}(\underline{k})}{\omega_{2Tj}^2 - \omega_m^2(\underline{k})} \quad , \tag{4-115}$$

$$Q_{3j}^{(m)}(\underline{k}) = \frac{s_3(n^2-\varepsilon_1)}{s_1(n^2-\varepsilon_3)} \cdot \frac{B_{3j}^{12} E_1^{(m)}(\underline{k})}{\omega_{3Tj}^2 - \omega_m^2(\underline{k})} \quad . \tag{4-116}$$

$E_1^{(m)}(\underline{k})$ can be regarded as a normalization factor and set = 1. $E_2^{(m)}(\underline{k})$ and $E_3^{(m)}(\underline{k})$ are correspondingly determined as a function of $E_1^{(m)}(\underline{k})$ by (4-112).

The equations (4-114) through (4-116) can be written in symmetric form by introducing the magnitude of the vector $\underline{E}^{(m)}(\underline{k})$. Taking (4-112) into account the square of the magnitude is

$$(E^{(m)})^2 = (E_1^{(m)})^2 + \frac{s_2^2(n^2-\varepsilon_1)}{s_1^2(n^2-\varepsilon_2)}(E_1^{(m)})^2 + \frac{s_3^2(n^2-\varepsilon_1)^2}{s_1^2(n^2-\varepsilon_3)^2}(E_1^{(m)})^2$$

$$= \frac{(n^2-\varepsilon_1)^2}{s_1^2}\left[\frac{s_1^2}{(n^2-\varepsilon_1)^2} + \frac{s_2^2}{(n^2-\varepsilon_2)^2} + \frac{s_3^2}{(n^2-\varepsilon_3)^2}\right](E_1^{(m)})^2$$

or simply

$$E_1^{(m)}(\underline{k}) = \frac{s_1}{(n^2-\varepsilon_1)} \cdot \frac{E^{(m)}(\underline{k})}{S(\underline{k})} \quad . \tag{4-117}$$

Hence

$$E_2^{(m)}(\underline{k}) = \frac{s_2}{(n^2-\varepsilon_2)} \cdot \frac{E^{(m)}(\underline{k})}{S(\underline{k})} \quad , \tag{4-118}$$

and

$$E_3^{(m)}(\underline{k}) = \frac{s_3}{(n^2-\varepsilon_3)} \cdot \frac{E^{(m)}(\underline{k})}{S(\underline{k})} \quad . \tag{4-119}$$

For simplicity, we used

$$S(\underline{k}) = \left(\frac{s_1^2}{(n^2-\varepsilon_1)^2} + \frac{s_2^2}{(n^2-\varepsilon_2)^2} + \frac{s_3^2}{(n^2-\varepsilon_3)^2}\right)^{1/2} \quad . \tag{4-120}$$

When finally we substitute (4-117) through (4-120) in (4-114) through (4-116), the quasinormal coordinates explicitly become

$$Q_{1j}^{(m)}(\underline{k}) = \frac{s_1}{(n^2-\varepsilon_1)} \cdot \frac{B_{1j}^{12}}{\omega_{1Tj}^2-\omega_m^2(\underline{k})} \cdot \frac{E^{(m)}(\underline{k})}{S(\underline{k})} \quad , \tag{4-121}$$

$$Q_{2j}^{(m)}(\underline{k}) = \frac{s_2}{(n^2-\varepsilon_2)} \cdot \frac{B_{2j}^{12}}{\omega_{2Tj}^2-\omega_m^2(\underline{k})} \cdot \frac{E^{(m)}(\underline{k})}{S(\underline{k})} \quad , \tag{4-122}$$

and

$$Q_{3j}^{(m)}(\underline{k}) = \frac{s_3}{(n^2-\varepsilon_3)} \cdot \frac{B_{3j}^{12}}{\omega_{3Tj}^2-\omega_m^2(\underline{k})} \cdot \frac{E^{(m)}(\underline{k})}{S(\underline{k})} \quad . \tag{4-123}$$

$E^{(m)}(\underline{k})/S(\underline{k})$ in the following can again be regarded as a normalization factor.

4.12 POLARITON SCATTERING INTENSITIES

Theoretical studies concerning Raman scattering intensities of polaritons are not only of interest for the spontaneous Raman effect but also in the search for suitable materials for stimulated Raman scattering experiments, see 5.1 and /74, 75, 162, 163/. Furthermore, intensity measurements form an elegant method for the determination of the electro-optic coefficients /97/. The earliest theoretical studies were published by Poulet /89/, Loudon /31/, Ovander /165/, and Burstein et al. /166/. Here we mainly follow the form used by Burstein.

The relative scattering intensity I is classically determined by the effective Raman tensor $\chi(\omega)$ in the following way

$$I \sim |\underline{e}_i \cdot \chi(\omega) \cdot \underline{e}_s|^2 \quad , \tag{4-124}$$

see (1-8) and 2.6. Herein $\underline{e}_i$ denotes a unit vector parallel to the (electric) polarization of the incident light wave and $\underline{e}_s$ a unit vector parallel to the polarization of the scattered wave.

The absolute scattering efficiency per unit solid angle becomes (for a Stokes process)

$$(dI/d\Omega) = (\omega_i/c)^4 VL|\langle n_\omega+1|\underline{e}_s \chi(\omega) \underline{e}_i|n_\omega\rangle|^2 \quad . \tag{4-125}$$

V is the scattering volume and L the travel length of an incident laser photon with the frequency ω_i in the material. The quantity inside the absolute value signs is the matrix element of the operator $\chi(\omega)$ between the state with n_ω polaritons of frequency ω and the state with $n_\omega+1$ polaritons. In view of the great difficulties experienced in conducting experimental studies designed to determine reliable absolute polariton scattering intensities, the information obtained from such data seems disappointing. Yet many still unknown properties of materials might easily be derived from the relative intensities. In spite of this, very few such data have been published in the literature. We know that this is because, even here, the experimental difficulties are still very great. A detailed discussion of (4-125) was given in the excellent article by Mills and Burstein /300/.

In the theory of Placzek /12/ polarizability has been treated as a function of the normal coordinates: $\alpha = \alpha(\underline{Q})$, see 1.2. For phonons in crystalline materials, α has to be replaced by the macroscopic susceptibility, $\chi = \chi(\underline{Q})$. In the polariton region, however, χ is to a large extent also determined by the electric field, $\chi = \chi(\underline{Q},\underline{E})$. The Placzek expansion presented in (1-10) therefore has to be performed with respect to the two variables $\underline{Q}$ and $\underline{E}$. In the 'linear' Raman effect the contributions of the normal coordinates and the electric field components are superimposed linearly; hence the Raman tensor elements for polaritons are given by

$$\chi_{\beta\gamma}^{(m)}(\omega) = \sum_j \sum_{\alpha=1}^{3} (\partial\chi_{\beta\gamma}/\partial Q_{\alpha j}) Q_{\alpha j}^{(m)}(\omega) + \sum_{\alpha=1}^{3} (\partial\chi_{\beta\gamma}/\partial E_\alpha) E_\alpha^{(m)}(\omega)$$

or

$$\chi_{\beta\gamma}^{(m)}(\omega) = \sum_j \sum_{\alpha=1}^{3} a_{\alpha,\beta\gamma}^{(j)} Q_{\alpha j}^{(m)}(\omega) + \sum_{\alpha=1}^{3} b_{\alpha,\beta\gamma} E_\alpha^{(m)}(\omega) \quad . \qquad (4\text{-}126)$$

The coefficients $a_{\alpha,\beta\gamma}^{(j)} = \partial\chi_{\beta\gamma}/\partial Q_{\alpha j}$ form the atomic displacement tensor and the $b_{\alpha,\beta\gamma} = \partial\chi_{\beta\gamma}/\partial E_\alpha$ the electro-optic tensor. When resonance Raman scattering is excluded, the coefficients $a_{\alpha,\beta\gamma}^{(j)}$ and $b_{\alpha,\beta\gamma}$ can be regarded as frequency-independent, see 1.2.

According to (4-126) the Raman tensor can be derived if we know

1) the quasinormal coordinates and the macroscopic electric field, and

2) the coefficients in (4-126), i.e. when the atomic displacement tensor and the electro-optic tensor have been numerically determined.

The quasinormal coordinates and the components of the electric field are obtained from (4-121) through (4-123) and (4-117) through (4-119). Because the components of the atomic displacement tensor cannot be numerically calculated from any microscopic theories so far known, they have necessarily to be determined experimentally. For this purpose it is sufficient to carry out intensity measurements concerning only the purely transverse, long-wavelength lattice vibrations for the principal directions of the crystal. For the k-th TO mode in α direction

$$\chi_{\alpha,\beta\gamma}^{(k)} = a_{\alpha,\beta\gamma}^{(k)} \quad , \qquad (4\text{-}127)$$

as can be seen from (4-126).
The electric field vanishes for TO modes with $k \approx 10^5\ cm^{-1}$ ($\underline{k} \to \infty$ limit, see 3.2). For fixed polarizations $\underline{e}_i$ and $\underline{e}_s$, the scattering intensity then becomes simply

$$I \sim (a^{(k)}_{\alpha,\beta\gamma})^2 \quad . \tag{4-128}$$

Except for the sign, the coefficients $a^{(k)}_{\alpha,\beta\gamma}$ can thus be determined from data on the relative scattering intensities of TO phonons only.

The components $b_{\alpha,\beta\gamma}$ of the electro-optic tensor, on the other hand, are related to those of the second harmonic generation tensor $d_{\alpha,\beta\gamma}$ by

$$b_{\alpha,\beta\gamma} = 4d_{\alpha,\beta\gamma} \quad . \tag{4-129}$$

They can therefore be determined (including sign) by well-known experimental methods from nonlinear optics. The signs of the $a^{(k)}_{\alpha,\beta\gamma}$, which according to (4-128) are still unknown, can be found by fitting the scattering intensities for arbitrary $\underline{k}$ directions and polarizations as, for instance, for the LO modes in the principal directions.

We now substitute the normal coordinates derived in 4.11 into (4-126), see also /159/. Taking into account (4-117) through (4-119) and (4-121) through (4-123), we obtain

$$\chi^{(m)}_{\beta\gamma} = \left[\sum_{\alpha=1}^{3} \frac{s_\alpha}{(n^2-\varepsilon_\alpha)} \left(b_{\alpha,\beta\gamma} + \sum_{j=1}^{n_\alpha} \frac{a^{(j)}_{\alpha,\beta\gamma}\ B^{12}_{\alpha j}}{\omega^2_{\alpha Tj} - \omega^2_m(\underline{k})}\right)\right] \frac{E^{(m)}(\underline{k})}{S(\underline{k})} \quad . \tag{4-130}$$

As pointed out before, the factor $E^{(m)}(\underline{k})/S(\underline{k})$ can be regarded as a normalization factor and set = 1. The frequencies $\omega_m(\underline{k})$ on the other hand are derived from Fresnel's equation.

Eq. (4-130) will now be discussed for some important special and limiting cases.

$k \to \infty$ requires $n^2 \to \infty$, see (3-20). From (4-120) it follows accordingly that $(n^2-\varepsilon_\alpha)S(\underline{k}) \to 1$. For long-wavelength optical phonons (4-130) thus reduces to

$$\chi^{(m)}_{\beta\gamma} = \left[\sum_{\alpha=1}^{3} s_\alpha \left(b_{\alpha,\beta\gamma} + \sum_{j=1}^{n_\alpha} \frac{a^{(j)}_{\alpha,\beta\gamma}\ B^{12}_{\alpha j}}{\omega^2_{\alpha Tj} - \omega^2_m(\underline{k})}\right)\right] E^{(m)}(\underline{k}) \quad . \tag{4-131}$$

This relation is essentially equivalent to an expression derived by Hartwig et al. /93/.

For ordinary polaritons in uniaxial crystals $n^2 = \varepsilon_1 = \varepsilon_2$ and $(n^2-\varepsilon_3)S(\underline{k}) \to \infty$ and $(n^2-\varepsilon_1)S(\underline{k}) = (n^2-\varepsilon_2)S(\underline{k}) \to \sqrt{s_1^2+s_2^2}$. Eq. (4-130) then becomes

$$\chi_{\beta\gamma}^{(m)} = \left[s_1 \left(b_{1,\beta\gamma} + \sum_{j=1}^{n_1} \frac{a_{1,\beta\gamma}^{(j)} B_{1j}^{12}}{\omega_{\perp j}^2 - \omega_m^2(\underline{k})} \right) + \right.$$

$$\left. + s_2 \left(b_{2,\beta\gamma} + \sum_{j=1}^{n_2} \frac{a_{2,\beta\gamma}^{(j)} B_{2j}^{12}}{\omega_{\perp j}^2 - \omega_m^2(\underline{k})} \right) \right] \frac{E^{(m)}(\underline{k})}{\sqrt{s_1^2+s_2^2}} \quad . \qquad (4\text{-}132)$$

The 1,2 plane is isotropic only with respect to the frequencies. The scattering intensities, however, remain direction-dependent.*)

For extraordinary polaritons, on the other hand, no essential simplification is obtained. The refractive index has to be calculated from

$$n^2 = \varepsilon_{\parallel}\varepsilon_{\perp}/(\varepsilon_{\perp}s_{\perp}^2 + \varepsilon_{\parallel}s_{\parallel}^2) \quad , \qquad (4\text{-}133)$$

see (4-37).

The scattering tensor for cubic crystals shows a similar shape to (4-132) for ordinary polaritons in uniaxial materials

$$\chi_{\beta\gamma}^{(m)} = \left[\sum_{\alpha=1}^{3} s_\alpha \left(b_{\alpha,\beta\gamma} + \sum_{j=1}^{r} \frac{a_{\alpha,\beta\gamma}^{(j)} B_{\alpha j}^{12}}{\omega_{Tj}^2 - \omega_m^2(\underline{k})} \right) \right] E^{(m)}(\underline{k}) \quad . \qquad (4\text{-}134)$$

The number of (threefold degenerate) modes has been indicated by $r \equiv n_1 = n_2 = n_3$.

If the crystal is piezoelectric and thus lacks an inversion center, only the coefficients $b_{3,12}$ and $a_{3,12}^{(j)}$ are different from zero (the indices can be arbitrarily permuted). In all other cases the Raman tensor vanishes and the polaritons are Raman-inactive.

Detailed experimental investigations of polariton scattering intensities are very rare. Systematic studies concerning the limiting case $\underline{k} \to \infty$ have been carried out by Unger and Schaack for $BeSO_4 \cdot H_2O$

*) Note that (4-132) has been corrected relative to /159/.

/95, 158/ and by Claus for $LiNbO_3$ /164/. Some typical results for $BeSO_4 \cdot H_2O$ are plotted in Fig.42. Data for α quartz have been published by Scott and Ushioda /215/, and for ZnSe by Ushioda et al. /301/.

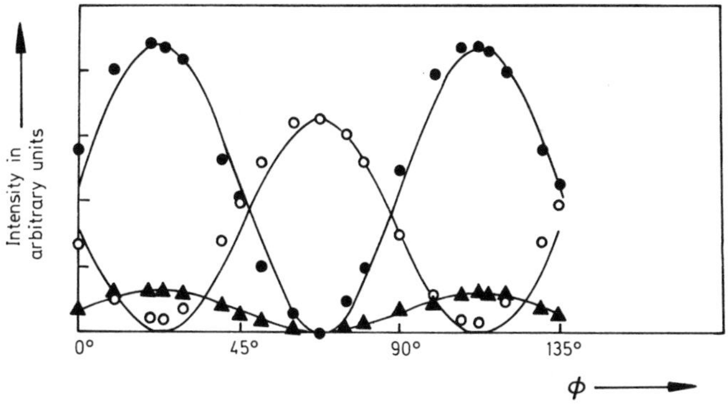

Fig.42 Calculated and experimentally recorded right-angle Raman scattering intensities from longitudinal and transverse phonons of type E in $BeSO_4 \cdot 4\ H_2O$ as a function of the <u>k</u> direction in the optically isotropic plane. ● = E(LO) at 1110 cm^{-1}, ▲ = E(LO) at 1162 cm^{-1} and o = E(TO) at 1084 cm^{-1} from /95/. ϕ denotes the angle between the x axis and the wave-vector of the incident laser radiation.

<u>Additional Literature</u>

Scott, J.E., Ushioda, S.: Polariton intensities in α quartz /215/.

Benson, H.J., Mills, D.L.: Polaritons in tetragonal $BaTiO_3$. Theory of the Raman lineshape /219/.

Otaguro, W.S., Wiener-Avnear, E., Porto, S.P.S.: Determination of second-harmonic generation coefficient and the linear electro-optic coefficient in $LiIO_3$ through oblique Raman phonon measurements /255/.

Scott, J.F., Damen, T.C., Shah, J.: Electro-optic and deformation potential contributions to the Raman tensor in CdS /281/.

Etchepare, J. Mathieu Merian: Calcul des intensités Raman des monocristaux de quartz α et β /282/.

Bendow, B.: Polariton theory of Raman scattering in insulating crystals /221/.

Agranovich, V.M., Ginzburg, V.L.: Theory of Raman scattering of light with formation of polaritons (real excitons) /235/.

Mavroyannis, C.: Optical excitation spectrum of interacting polariton waves in molecular crystals /255/.
Wemple, H.: Elektro-optic phenomena in crystals /312/.

4.13 THE MICROSCOPIC THEORY

The two first fundamental equations of the polariton theory (4-7) and (4-8) were derived in 4.1 from the energy density (4-4). The sublattices have been regarded as rigid. This assumption caused the coefficients $B^{\mu\nu}$ to be constant, i.e. independent of $\underline{k}$ and ω. We now show that the two fundamental equations in question can also be derived from a microscopic model and that the coefficients $B^{\mu\nu}$ can be shown to be constant in the long-wavelength limit. The constancy in general is well fulfilled for $k \leq 10^7\ \mathrm{cm}^{-1}$. Thus, lattice dynamics provides a microscopic interpretation of the coefficients, i.e. a method for calculating them from microscopic data on the materials. A discussion of the corresponding simplest model was included in 3.1. The third fundamental equation of the polariton theory, (4-9) is left untouched because it was been derived from Maxwell's equations and does not contain any material properties.

The model considered by Born and v. Kármán, see 3.1, cannot be used to derive the fundamental equations of the polariton theory because long-range Coulomb forces have not been taken into account. As a result the coefficients B^{12} and B^{21} vanish. We therefore use the rigid-ion model introduced by Kellermann /263/, see also /62/.

An atom in its equilibrium position is represented by the position vector $\underline{R}_{\ell,k} = \underline{R}_\ell+\underline{R}_k$. When it is displaced, its actual position is indicated by $\underline{r}_{\ell,k} = \underline{R}_{\ell,k}+\underline{u}_{\ell,k}$, see Fig.43. The displacement from the equilibrium position is described by the vector $\underline{u}_{\ell,k}$. $\underline{R}_\ell$ is the position vector of a certain fixed point in the unit cell (in general, a corner) and $\underline{R}_k$ describes the distance from this point to the atom. The index ℓ denotes the unit cell and $k = 1,\ldots,n$ counts the number of atoms in the cell.

In the adiabatic approximation a lattice potential Φ exists as a function of $\underline{r}_{\ell,k}$ or $\underline{u}_{\ell,k}$. The motion of an atom (ℓ,k) is therefore described by the equation

Fig.43 The rigid-ion model of Kellermann, see text.

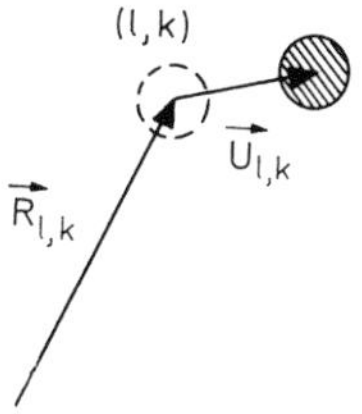

$$m_k \ddot{\underline{u}}_{\ell,k} = - \frac{\partial \Phi}{\partial \underline{u}_{\ell,k}} = - \left[\left(\frac{\partial \Phi}{\partial \underline{u}_{\ell,k}} \right)_o + \sum_{\ell',k'} \left(\frac{\partial^2 \Phi}{\partial \underline{u}_{\ell,k} \partial \underline{u}_{\ell',k'}} \right)_o \underline{u}_{\ell',k'} + \ldots \right] \quad . \tag{4-135}$$

Φ is expanded in the equilibrium in terms of $\underline{u}_{\ell,k}$. m_k denotes the mass of the k-th atom. We use common abbreviations for the coefficients /62/

$$\Phi_{\ell,k} \equiv \left(\frac{\partial \Phi}{\partial \underline{u}_{\ell,k}} \right)_o \quad \text{and} \quad \Phi_{\substack{\ell,\ell' \\ k,k'}} \equiv \left(\frac{\partial^2 \Phi}{\partial \underline{u}_{\ell,k} \partial \underline{u}_{\ell',k'}} \right)_o \quad . \tag{4-136}$$

The latter are usually referred to as force constants. By taking only the quadratic terms of the expansion into account, we obtain the equation of motion in the harmonic approximation

$$m \, \ddot{\underline{u}}_{\ell,k} = - \sum_{\ell',k'} \Phi_{\substack{\ell,\ell' \\ k,k'}} \cdot \underline{u}_{\ell',k'} \quad . \tag{4-137}$$

The constant term vanishes in the equilibrium and the terms $-\Phi_{\substack{\ell,\ell' \\ k,k'}} \cdot \underline{u}_{\ell',k'}$ thus describe the additional forces with which the atoms (ℓ',k') act on the atom (ℓ,k) when only the former are displaced. If the crystal is regarded as infinite, (4-137) determines an infinite system of linear differential equations which may be solved with a plane-wave ansatz

$$\underline{u}_{\ell,k} = \underline{U}_k \exp(i \, \underline{k} \cdot \underline{R}_{\ell,k} - i\omega t) \quad . \tag{4-138}$$

The periodicity of the crystal has been taken into account by assuming the amplitudes to be equal in equivalent lattice points.

By combining (4-137) and (4-138), we obtain

$$m_k \omega^2 \underline{U}_k = \sum_{\ell',k'} \Phi_{\substack{\ell,\ell' \\ k,k'}} \underline{U}_{k'} \exp[i\ \underline{k}\cdot(\underline{R}_{\ell',k'} - \underline{R}_{\ell,k})]$$

$$= \sum_{k'} C_{k,k'}(\underline{k}) \cdot \underline{U}_{k'} \quad . \tag{4-139}$$

Here

$$C_{k,k'}(\underline{k}) \equiv \sum_{\ell'} \Phi_{\substack{\ell,\ell' \\ k,k'}} \exp(-i\ \underline{k}\cdot\underline{R}_{\substack{\ell,\ell' \\ k,k'}}) \tag{4-140}$$

and

$$\underline{R}_{\substack{\ell,\ell' \\ k,k'}} \equiv \underline{R}_{\ell,k} - \underline{R}_{\ell',k'} \quad . \tag{4-141}$$

$C_{k,k'}(\underline{k})$ has become known as the dynamical matrix, which is independent of the cell index ℓ because the force constants $\Phi_{\substack{\ell,\ell' \\ k,k'}}$ and the distance vectors $\underline{R}_{\substack{\ell,\ell' \\ k,k'}}$ depend only on the difference in the cell indices $\ell-\ell'$. We therefore also use the nomenclature $\Phi_{\substack{\ell-\ell' \\ k,k'}}$ and $\underline{R}_{\substack{\ell-\ell' \\ k,k'}}$.

The equations have been reduced to a set of linear homogeneous vector equations whose number is equal to the number of atoms in the unit cell. As long as no coupling with electromagnetic waves takes place, the dispersion branches $\omega = \omega(\underline{k})$ are determined by the condition that the determinant of (4-139) has to vanish

$$|C_{k,k'}(\underline{k}) - m_k \omega^2 \delta_{k,k'} I| = 0 \quad . \tag{4-142}$$

I denotes the unit matrix. The number of branches, which is identical with the number of solutions, is 3n.

We now turn our attention to ionic crystals where long-range Coulomb forces act in addition to the short-range forces. As first shown by Kellermann /263/, a term describing the macroscopic field can be separated from (4-139) for such crystals. In order to show this, we formally split $C_{k,k'}(\underline{k})$ into two parts, $C^N_{k,k'}(\underline{k})$ describing the short-range forces, and $C^C_{k,k'}(\underline{k})$ describing the Coulomb forces:

$$C_{k,k'}(\underline{k}) = C^N_{k,k'}(\underline{k}) + C^C_{k,k'}(\underline{k}) \quad . \tag{4-143}$$

Because the force constants $\Phi_{\substack{\ell,\ell' \\ k,k'}}$ rapidly attenuate with distance from the atom (ℓ,k), it is sufficient to take into account very few neighbors in $C^N_{k,k'}(\underline{k})$ (e.g. second or third neighbors). The Coulomb forces on the other hand must be examined somewhat more carefully. The contribution of the Coulomb potential to Φ is

$$\Phi^C = (1/2) \sum_{\substack{\ell,\ell' \\ k,k'}}{}' e_k e_{k'} / |\underline{R}_{\ell,k} - \underline{R}_{\ell',k'}| \quad , \tag{4-144}$$

where e_k and $e_{k'}$ are the effective ionic charges of the sublattices k and k', respectively. The upper prime on the sum symbol excludes the term with $\ell = \ell'$ and $k = k'$. Taking (4-136), (4-140) and (4-143) into account, we obtain

$$\begin{aligned} C^C_{k,k'}(\underline{k}) &= \sum_{\ell}{}' \frac{\partial^2}{\partial \underline{R}_{\ell,k} \partial \underline{R}_{k'}} \left(\frac{e_k e_{k'}}{|\underline{R}_{\ell,k} - \underline{R}_{k'}|} \right)_0 \exp[-i\, \underline{k}\cdot(\underline{R}_{\ell,k} - \underline{R}_{k'})] \\ &= -e_k e_{k'} \exp(i\, \underline{k}\cdot\underline{R}_{k'}) \sum_{\ell}{}' \frac{\partial^2}{\partial \underline{R}_{k'} \partial \underline{R}_{k'}} \left(\frac{\exp(-i\, \underline{k}\cdot\underline{R}_{\ell,k})}{|\underline{R}_{\ell,k} - \underline{R}_{k'}|} \right) . \end{aligned} \tag{4-145}$$

If, for simplicity, an effective field is introduced

$$\underline{F}_k(\underline{k}) = -(1/e_k) \sum_{k'} C^C_{k,k'}(\underline{k}) \cdot \underline{U}_{k'} \quad , \tag{4-146}$$

(4-139) can be written

$$m_k \omega^2 \underline{U}_k = \sum_{k'} C^N_{k,k'}(\underline{k}) \cdot \underline{U}_{k'} - e_k \underline{F}_k(\underline{k}) \quad . \tag{4-147}$$

$\underline{F}_k(\underline{k})$ denotes the effective local electric field acting at the position of a certain atom in the sublattice k. We later show in detail (Appdx 1) that the effective field can be split into a part independent of the sublattice and another part depending on the sublattice

$$\underline{F}_k(\underline{k}) = \underline{E}(\underline{k}) - \sum_{k'} (1/e_k) B_{k,k'}(\underline{k}) \cdot \underline{U}_{k'} \quad . \tag{4-148}$$

The k-independent part $\underline{E}(\underline{k})$ obviously describes the macroscopic electric field. The $B_{k,k'}(\underline{k})$ terms on the other hand behave like the terms for short-range forces, $C^N_{k,k'}(\underline{k})$. By introducing the amplitude of the dipole moment $\underline{p}_k = e_k \underline{U}_k$ and defining the contribution of the sublattice k to the macroscopic polarization as $\underline{P}_k = \underline{p}_k/\upsilon_a$, we can write (4-148)

$$\underline{F}_k(\underline{k}) = \underline{E}(\underline{k})+\sum_{k'}\gamma_{k,k'}(\underline{k})\cdot\underline{P}_{k'} \quad . \tag{4-149}$$

$\gamma_{k,k'}(\underline{k}) = -\upsilon_a B_{k,k'}(\underline{k})/e_k e_{k'}$ is a generalized Lorentz tensor (or simply Lorentz factor). On substituting (4-149) into the equation of motion (4-147), we have

$$-m_k\omega^2\underline{U}_k = -\sum_{k'}C^G_{k,k'}(\underline{k})\cdot\underline{U}_{k'}+e_k\underline{E}(\underline{k}) \tag{4-150}$$

where

$$C^G_{k,k'}(\underline{k}) = C^N_{k,k'}(\underline{k})+B_{k,k'}(\underline{k}) \quad . \tag{4-151}$$

We furthermore introduce the sublattice densities $\rho_k = m_k/\upsilon_a$ into (4-150)

$$-\rho_k\omega^2\underline{U}_k = -\sum_{k'}(1/\upsilon_a)C^G_{k,k'}(\underline{k})\cdot\underline{U}_k+(e_k/\upsilon_a)\underline{E}(\underline{k}) \quad . \tag{4-152}$$

Eq.(4-152) can be written as a vector equation by using the abbreviations

$$L_{k,k'}(\underline{k}) \equiv C^G_{k,k'}(\underline{k})/\upsilon_a$$

and (4-153)

$$M_{k\alpha} \equiv e_k/\upsilon_a \quad .$$

Finally, a matrix ρ containing the sublattice densities as diagonal elements is introduced in the same way as in 4.1. The $\underline{U}_k$ then form a (3n-3)-dimensional vector $\underline{U}$ and (4-152) takes the form

$$-\omega^2\rho\underline{U} = -L(\underline{k})\cdot\underline{U} + M\cdot\underline{E} \quad . \tag{4-154}$$

This relation corresponds to (4-2) except for the $\underline{k}$ dependence of L. We later show that $L(\underline{k})$ is almost constant in the region $k \leq 10^7$ cm^{-1}. Because the transformation to normal coordinates can be performed in the same way as in 4.1, (4-154) is identical with the first fundamental equation of the polariton theory

$$-\omega^2\underline{Q} = B^{11}(\underline{k})\cdot\underline{Q} + B^{12}\cdot\underline{E} \quad . \tag{4-155}$$

The macroscopic polarization, too, can easily be expressed by means of the rigid-ion model. Obviously

$$\underline{P} = \sum_k \underline{p}_k/\upsilon_a = \sum_k e_k\underline{U}_k/\upsilon_a = \sum_k M_{k\alpha}\underline{U}_k \tag{4-156}$$

or, in matrix form,

$$\underline{P} = M^+ \cdot \underline{U} \quad . \tag{4-157}$$

When the vector $\underline{Q}$ of the normal coordinates is introduced instead of the displacement vector $\underline{U}$, the equation can be written

$$\underline{P} = B^{21} \cdot \underline{Q} \quad . \tag{4-158}$$

Eq.(4-158) is identical with the second fundamental equation (4-8), except that the term $B^{22}\cdot\underline{E}$ does not appear. This, however, is a specific characteristic of the rigid-ion model. $B^{22}\cdot\underline{E}$ describes the electronic polarization, which is nonexistent for rigid-ions. If we take into account electronic polarization in addition, the microscopic theory yields the complete set of fundamental equations (4-7) and (4-8), including the term $B^{22}\cdot\underline{E}$. Such models are e.g. the shell model /264-267/, its generalizations /268, 269/, and the bond-charge model /270-275/. Other microscopic models, as discussed by, for instance, Pick /80/ and in /276/ and /277/, also allow derivation of the fundamental equations in the long-wavelength limit, see Appdx 2. For review articles see refs /355-357/.

We finally show the constancy of the coefficients in the long-wavelength limit. In the rigid-ion model obviously only $L_{k,k'}(\underline{k}) = (1/\upsilon_a)C^G_{k,k'}(\underline{k})$ is a function of $\underline{k}$,see (4-152) and (4-140). We first consider the contribution $C^N_{k,k'}(\underline{k})$ describing the short-range forces, (4-143). Terms with the smallest distance vectors $\underline{R}_{\ell,\ell' \atop k,k'}$ provide the main contribution to $C^N_{k,k'}(\underline{k})$ because the forces and hence the force constants $\Phi_{\ell,\ell' \atop k,k'}$ rapidly attenuate for increasing distances. The contributions of the short-range forces are thus constant in good approximation when the magnitudes of the exponents of the exponential factors, (4-140), are very small, i.e.

$$|\underline{k}\cdot\underline{R}_{\ell,\ell' \atop k,k'}| \leq k\, R_{\ell,\ell' \atop k,k'} \ll 1 \quad . \tag{4-159}$$

As a realistic value for the distance $\underline{R}_{\ell,\ell' \atop k,k'}$ of next neighbors we assume 0.25 nm = 2.5×10^{-8} cm . (For ZnS, e.g.,it is 0.234 nm.) The condition (4-159) therefore requires

$$k \ll (1/2.5) \cdot 10^{8}\ \mathrm{cm}^{-1} = 4 \cdot 10^{7}\ \mathrm{cm}^{-1} \quad . \tag{4-160}$$

For neighbors at greater distance the limit for the constancy of the corresponding terms is moved toward smaller k values. Since, however, the force constants $\Phi_{\ell,\ell' \atop k,k'}$ rapidly decrease with increasing distances, these terms involve only small corrections.

In lattice-dynamical calculations for real crystals, the short-range forces have generally been considered only for at maximum third neighbors. Looking again at the ZnS lattice,we see that third neighbors have barely twice the distance of next neighbors (0.448 nm). The condition (4-160) then alternatively becomes $k \ll 2 \times 10^{7}\ \mathrm{cm}^{-1}$. Consequently the $C^{N}_{k,k}(\underline{k})$ can be regarded as generally constant for $k \ll 10^{7}\ \mathrm{cm}^{-1}$.

As shown in Appdx 1, an analogous result is obtained for the $B_{k,k'}(\underline{k})$ orginating from Coulomb forces. According to (4-151) and (4-153), the constancy has therefore been verified also for the $C^{G}_{k,k'}(\underline{k})$ and the $L_{k,k'}(\underline{k})$. This implies that the coefficients B^{11}_{ij}, too, are constant in the region $k \ll 10^{7}\ \mathrm{cm}^{-1}$ because they can be expressed as linear combinations of the $C^{G}_{k,k'}$ or $L_{k,k'}$.

In the improved microscopic models that include electronic polarizability, the tensors B^{12}, B^{21}, and B^{22} again remain $\underline{k}$-independent: the $\underline{k}$ dependence is described by exponential functions essentially of the types cited above, (4-139), (4-140), and (4-145). The derivations presented can therefore easily be extended to these more complicated models.

The third fundamental equation of the polariton theory (4-9) is not affected because it was derived from Maxwell's equations.

4.14 POLARITONS AS PARTICLES

All properties of polaritons so far described have been theoretically derived by means of the wave model. We have seen that most experimental results of any importance can be correctly described by this classic model. In fact, the properties derived from the wave model and the particle model are to a large extent identical for stationary states. Quantization of the polariton waves, however, becomes of importance when we consider the mechanism of interaction processes, i.e. processes in which polaritons are created or destroyed. The Raman effect beautifully illustrates such processes: polaritons are created and destroyed, respectively, for Stokes and Anti-Stokes scattering. Interaction processes with polaritons have relatively rarely been discussed in the literature, and we want to outline polariton quantization only in principle. We restrict our discussion to cubic crystals with only one (threefold degenerate) infrared-active dispersion branch /2, 303, 324-327, 354/.

Electronic polarization is neglected because it is not of principal importance for the creation of a polariton, i.e. we define $\varepsilon_\infty = 1$ as requiring $B^{22} = 0$. For the microscopic model this implies a restriction to the rigid-ion model of Kellermann. The polarization of the lattice then originates only from the displacement of the rigid-ions, see 4.13.

In order to quantize the Hamilton function of polaritons, we discuss the energy density (3-34) in somewhat more detail for cubic crystals. The normal coordinates of LO modes can be omitted because they do not couple with electromagnetic waves. The fourth term in (3-34) vanishes ($B^{22} = 0$) and the three first terms are referred to as

$$H_P = (1/2)\dot{\underline{Q}}^2 - (1/2)B^{11}\underline{Q}^2 - B^{12}\underline{Q}\cdot\underline{E} \quad . \tag{4-161}$$

The last two terms are similarly referred to as

$$H_L = (1/8\pi)(\underline{E}^2 + \underline{H}^2) \quad . \tag{4-162}$$

H_P and H_L, respectively, denote the polarization energy of the medium and the electromagnetic energy in vacuum. If we neglect the interaction term $H_I = \underline{E}\cdot\underline{P}$, H_P and H_L are independent of each other.

They may thus be quantized separately. As is well known, the quanta of H_L are photons whereas those of H_P are referred to as polarization quants or optical phonons. In order to introduce the latter we rewrite H_P in terms of the macroscopic polarization, which in our approximation is simply $\underline{P} = B^{21}\underline{Q}$, see (4-6)

$$H_P = (1/2(B^{12})^2)(\dot{\underline{P}}^2 - B^{11}\underline{P}^2 + 4\pi(B^{12})^2\underline{P}^2) = (1/2\rho\omega_T^2)(\dot{\underline{P}}^2 + \omega_P^2\underline{P}^2) \quad . \tag{4-163}$$

Here

$$\omega_P^2 = -B^{11} + 4\pi(B^{12})^2 = \omega_T^2 + (\varepsilon_o - 1)\omega_T^2 = \varepsilon_o\omega_T^2 \tag{4-164}$$

and $4\pi\rho$ is the oscillator strength defined by (3-29). The frequency of the transverse phonon in the short-wavelength limit $k \to \infty$ outside the polariton region is denoted by ω_T. We have made use of the fact that the $\underline{E}$ field is electrostatic and thus $\underline{E} = -4\pi\underline{P}$. Expressed in terms of $\underline{Q}$, on the other hand, H_P takes the well-known form

$$H_P = (1/2)(\dot{\underline{Q}}^2 + \omega_P^2\underline{Q}^2) \tag{4-165}$$

which can easily be verified by substituting $\underline{P} = B^{21}\underline{Q}$. Finally, in order to demonstrate the connection with the derivation given by Hopfield /2, 303/, we rewrite the equation of motion (4-5)

$$\ddot{\underline{Q}} = B^{11}\underline{Q} + B^{12}\underline{E} \tag{4-166}$$

into an equation for $\underline{P}$. If we multiply (4-166) by B^{21} $(= B^{12})$ and again introduce $\underline{P} = B^{21}\underline{Q}$, we get

$$\ddot{\underline{P}} - B^{11}\underline{P} = (B^{12})^2\underline{E}$$

or explicitly

$$\ddot{\underline{P}} + \omega_T^2\underline{P} = (\varepsilon_o - 1)\omega_T^2\underline{E}/4\pi = \rho\omega_T^2\underline{E} \quad . \tag{4-167}$$

This relation is identical to that in Hopfield's paper /Ref.2,Eq.(4)/. Our ρ there is simply $\beta(\equiv\rho)$.

Creation and annihilation operators for optical phonons $a^+_{\underline{k}_i}$ and $a_{\underline{k}_i}$, respectively, are introduced in the normal way. $a^+_{\underline{k}_i}$ shows the following characteristics:

Acting on a system of optical phonons

$$|n_{\underline{k}_1}, n_{\underline{k}_2}, \ldots, n_{\underline{k}_i}, \ldots, n_{\underline{k}_N}>$$

an additional quant with the momentum $\hbar \underline{k}_i$ is created

$$a^+_{\underline{k}_i} |n_{\underline{k}_1}, n_{\underline{k}_2}, \ldots, n_{\underline{k}_i}, \ldots, n_{\underline{k}_N}> =$$

$$= \sqrt{n_{\underline{k}_i}+1}\, |n_{\underline{k}_1}, n_{\underline{k}_2}, \ldots, n_{\underline{k}_i}+1, \ldots, n_{\underline{k}_N}> \quad . \qquad (4\text{-}168)$$

The annihilation operator on the contrary destroys a phonon with the momentum $\hbar \underline{k}_i$

$$a_{\underline{k}_i} |n_{\underline{k}_1}, n_{\underline{k}_2}, \ldots, n_{\underline{k}_i}, \ldots, n_{\underline{k}_N}> =$$

$$= \sqrt{n_{\underline{k}_i}}\, |n_{\underline{k}_1}, n_{\underline{k}_2}, \ldots, n_{\underline{k}_i}-1, \ldots, n_{\underline{k}_N}> \quad . \qquad (4\text{-}169)$$

$\sqrt{n_{\underline{k}_i}+1}$ and $\sqrt{n_{\underline{k}_i}}$ are normalization factors and the discrete vectors $\underline{k}_i$ $(i = 1,\ldots,N)$ are due to the periodic boundary conditions of the finite space volume, which is here assumed to be a unit volume.

Some important properties of the operators $a^+_{\underline{k}_i}$ and $a_{\underline{k}_i}$ are summarized below. The reader is referred to /5, 325, 326, 354/ for detailed discussions.

The operator $\hat{n}_{\underline{k}_i} = a^+_{\underline{k}_i} a_{\underline{k}_i}$ determines the number of optical phonons with a momentum $\hbar \underline{k}_i$ (quantum number). According to (4-168) and (4-169), its eigenvalue is the quantum number

$$a^+_{\underline{k}_i} a_{\underline{k}_i} |n_{\underline{k}_1}, n_{\underline{k}_2}, \ldots, n_{\underline{k}_i}, \ldots, n_{\underline{k}_N}> =$$

$$a^+_{\underline{k}_i} \sqrt{n_{\underline{k}_i}}\, |n_{\underline{k}_1}, n_{\underline{k}_2}, \ldots, n_{\underline{k}_i}-1, \ldots, n_{\underline{k}_N}> =$$

$$\sqrt{(n_{\underline{k}_i}-1)+1}\; \sqrt{n_{\underline{k}_i}}\, |n_{\underline{k}_1}, n_{\underline{k}_2}, \ldots, n_{\underline{k}_i}, \ldots, n_{\underline{k}_N}> =$$

$$n_{\underline{k}_i} |n_{\underline{k}_1}, n_{\underline{k}_2}, \ldots, n_{\underline{k}_i}, \ldots, n_{\underline{k}_N}> \quad . \qquad (4\text{-}170)$$

The eigenvalue of the operator $a_{\underline{k}_i} a^+_{\underline{k}_i}$ correspondingly becomes

$(n_{\underline{k}_i}+1)$ and consequently the commutator

$$\left[a_{\underline{k}_i}, a^+_{\underline{k}_i}\right] = a_{\underline{k}_i} a^+_{\underline{k}_i} - a^+_{\underline{k}_i} a_{\underline{k}_i} = 1 \quad . \tag{4-171}$$

The commutator vanishes for $\underline{k}_i \neq \underline{k}_j$ and can therefore generally be written as

$$\left[a_{\underline{k}_j}, a^+_{\underline{k}_i}\right] = \delta_{ji} \quad . \tag{4-172}$$

The displacement vector $\underline{Q}$ (or alternatively the polarization $\underline{P}$) may be expressed by a Fourier series covering the unit volume

$$\underline{Q} = \sum_{\substack{\underline{k}_i \\ \lambda=1,2}} Q_o \underline{e}_{\underline{k}_i,\lambda} \left(a_{\underline{k}_i,\lambda} e^{i\,\underline{k}_i \cdot \underline{r}} + a^*_{\underline{k}_i,\lambda} e^{-i\,\underline{k}_i \cdot \underline{r}} \right) \quad . \tag{4-173}$$

$\lambda = 1,2$ refers to the two directions of the transverse polarization perpendicular to the wave-vector. These directions are described by the unit vectors $\underline{e}_{\underline{k}_i,\lambda}$. Longitudinal waves can be left out of consideration, as pointed out above.

The operator corresponding to (4-173) is obtained by simply replacing the amplitudes $a_{\underline{k}_i,\lambda}$ and $a^*_{\underline{k}_i,\lambda}$ by the annihilation and creation operators, respectively:

$$\underline{Q} = Q_o \sum_{\substack{\underline{k}_i \\ \lambda=1,2}} \underline{e}_{\underline{k}_i,\lambda} \left(a_{\underline{k}_i,\lambda} e^{i\,\underline{k}_i \cdot \underline{r}} + a^+_{\underline{k}_i,\lambda} e^{-i\,\underline{k}_i \cdot \underline{r}} \right) =$$

$$= Q_o \sum_{\substack{\underline{k}_i \\ \lambda=1,2}} \underline{e}_{\underline{k}_i,\lambda} \left(a_{\underline{k}_i,\lambda} + a^+_{-\underline{k}_i,\lambda} \right) e^{i\,\underline{k}_i \cdot \underline{r}} \tag{4-174}$$

where $Q_o = \sqrt{\hbar/2\omega_p}$. The operator of the canonical conjugate momentum $\dot{\underline{Q}}$ becomes

$$\dot{\underline{Q}} = -i\sqrt{\hbar\omega_p/2} \sum_{\substack{\underline{k}_i \\ \lambda=1,2}} \underline{e}_{\underline{k}_i,\lambda} \left(a_{\underline{k}_i,\lambda} - a^+_{-\underline{k}_i,\lambda} \right) e^{i\,\underline{k}_i \cdot \underline{r}} \quad . \tag{4-175}$$

The operators $\underline{P}$ and $\dot{\underline{P}}$ for the polarization*) are derived from (4-175) by multiplication by B^{21}

$$\underline{P} = B^{21}\sqrt{\hbar/2\omega_P} \sum_{\substack{\underline{k}_i \\ \lambda=1,2}} \underline{e}_{\underline{k}_i,\lambda}\left(a_{\underline{k}_i,\lambda} + a^+_{-\underline{k}_i,\lambda}\right) e^{i\,\underline{k}_i\cdot\underline{r}} \tag{4-176}$$

and

$$\dot{\underline{P}} = -iB^{21}\sqrt{\hbar\omega_P/2} \sum_{\substack{\underline{k}_i \\ \lambda=1,2}} \underline{e}_{\underline{k}_i,\lambda}\left(a_{\underline{k}_i,\lambda} - a^+_{-\underline{k}_i,\lambda}\right) e^{i\,\underline{k}_i\cdot\underline{r}} \quad . \tag{4-177}$$

If we introduce the operators $\underline{Q}$, $\dot{\underline{Q}}$ or $\underline{P}$, $\dot{\underline{P}}$ into (4-163) and (4-165) and take into account the commutation rules (4-172), the Hamilton operator for polar phonons becomes

$$H_P = \sum_{\substack{\underline{k}_i \\ \lambda=1,2}} \hbar\omega_P(a^+_{\underline{k}_i,\lambda}a_{\underline{k}_i,\lambda} + 1/2) = \sum_{\substack{\underline{k}_i \\ \lambda=1,2}} \hbar\omega_P(n_{\underline{k}_i,\lambda} + 1/2) \quad . \tag{4-178}$$

This result can be physically interpreted in the following way: neglecting the interaction energy $H_I = \underline{E}\cdot\underline{P}$, the system possesses $n_{\underline{k}_i,\lambda}$ phonons polarized in λ direction with the energy $\hbar\omega_P$ and wave-vectors $\underline{k}_i$. The phonon energy thus does not depend on the wave-vector in the uncoupled state.

The vacuum energy of the electromagnetic waves (4-162) can be quantized in an analogous way. We introduce creation and annihilation operators for photons $b^+_{\underline{k}_i}$ and $b_{\underline{k}_i}$, respectively, by analogy with (4-168) and (4-169), and replace $\underline{E}$ and $\underline{H}$ by the vector potential $\underline{A}$

$$\underline{E} = -(1/c)\dot{\underline{A}} \quad ; \quad \underline{H} = \mathrm{curl}\,\underline{A} \quad ; \quad \mathrm{div}\,\underline{A} = 0 \quad . \tag{4-179}$$

$\underline{A}$ can then be expressed by a Fourier series over the unit volume

$$\underline{A} = A_o \sum_{\substack{\underline{k}_i \\ \lambda=1,2}} \underline{e}_{\underline{k}_i,\lambda}\left(b_{\underline{k}_i,\lambda} e^{i\,\underline{k}_i\cdot\underline{r}} + b^*_{\underline{k}_i,\lambda} e^{-i\,\underline{k}_i\cdot\underline{r}}\right) \quad . \tag{4-180}$$

---

*) Note that $\underline{P}$ here denotes the polarization and not the canonical conjugate momentum $\dot{\underline{Q}}$ of $\underline{Q}$ as in most textbooks.

The corresponding operator is again obtained by replacing the amplitudes by creation and annihilation operators. Thus, we get

$$\underline{A} = A_o \sum_{\substack{\underline{k}_i \\ \lambda=1,2}} \underline{e}_{\underline{k}_i,\lambda}\left(b_{\underline{k}_i,\lambda}+b^+_{-\underline{k}_i,\lambda}\right)e^{i\,\underline{k}_i\cdot\underline{r}} \quad , \tag{4-181}$$

with $A_o = \sqrt{2\pi c/\hbar k_i}$. Apart from a constant factor, $\underline{E}$ can be interpreted as the canonical conjugate operator of $\underline{A}$

$$\underline{E} = -(1/c)\dot{\underline{A}} = E_o \sum_{\substack{\underline{k}_i \\ \lambda=1,2}} \underline{e}_{\underline{k}_i,\lambda}\left(b_{\underline{k}_i,\lambda}-b^+_{-\underline{k}_i,\lambda}\right)e^{i\,\underline{k}_i\cdot\underline{r}} \quad , \tag{4-182}$$

where

$$E_o = -i\sqrt{2\pi\hbar c k_i} \quad . \tag{4-183}$$

By analogy with (4-178), we finally obtain

$$H_L = \sum_{\substack{\underline{k}_i \\ \lambda=1,2}} \hbar\omega_L(\underline{k}_i)(b^+_{\underline{k}_i,\lambda}b_{\underline{k}_i,\lambda}+1) = \sum_{\substack{\underline{k}_i \\ \lambda=1,2}} \hbar\omega_L(\underline{k}_i)(n_{\underline{k}_i,\lambda}+1) \quad , \tag{4-184}$$

where

$$\omega_L(k_i) = ck_i \quad . \tag{4-185}$$

Thus the electromagnetic field in the uncoupled state contains $n_{\underline{k}_i,\lambda}$ photons with the energy $\hbar c k_i$ polarized in λ direction. For more detailed discussions, we refer to the textbooks by Heitler /327/ and Louisell /331/.

We now include the interaction term $H_I = \underline{E}\cdot\underline{P}$. Taking into account (4-176) and (4-182) for $\underline{P}$ and $\underline{E}$, the operator becomes

$$H_I = i\hbar D \sum_{\substack{\underline{k}_i \\ \lambda=1,2}} (a_{\underline{k}_i,\lambda}+a^+_{-\underline{k}_i,\lambda})(b_{-\underline{k}_i,\lambda}-b^+_{\underline{k}_i,\lambda}) \tag{4-186}$$

where

$$D = B^{21}(\pi c k_i/\omega_P)^{1/2} = (\pi\rho\omega_T^2 c k_i/\omega_P)^{1/2} \quad . \tag{4-187}$$

The Hamilton operator for the coupled system is therefore

$$H = \sum_{\substack{\underline{k}_i \\ \lambda=1,2}} \Big[\hbar\omega_P(a^+_{\underline{k}_i,\lambda}a_{\underline{k}_i,\lambda}+1/2)+\hbar\omega_L(b^+_{\underline{k}_i,\lambda}b_{\underline{k}_i,\lambda}+1/2) +$$

$$-i\hbar D(a^+_{\underline{k}_i,\lambda}b^+_{-\underline{k}_i,\lambda}-a^+_{\underline{k}_i,\lambda}b_{\underline{k}_i,\lambda}+a_{-\underline{k}_i,\lambda}b^+_{-\underline{k}_i,\lambda}-a_{-\underline{k}_i,\lambda}b_{\underline{k}_i,\lambda})\Big] \quad . \tag{4-188}$$

Because of the coupling term, phonons and photons are no longer eigenstates of this system. New operators $\alpha^{(1)+}_{\underline{k}_i,\lambda}$, $\alpha^{(1)}_{-\underline{k}_i,\lambda}$, $\alpha^{(2)+}_{\underline{k}_i,\lambda}$, and $\alpha^{(2)}_{-\underline{k}_i,\lambda}$, however, may be defined by linear combinations of the $a^+_{\underline{k}_i,\lambda}$, $a_{-\underline{k}_i,\lambda}$, $b^+_{\underline{k}_i,\lambda}$, and $b_{-\underline{k}_i,\lambda}$ in the following way:

$$\alpha^{(1)+}_{\underline{k}_i,\lambda} = C_{11}(\underline{k}_i)a^+_{\underline{k}_i,\lambda}+C_{12}(\underline{k}_i)a_{-\underline{k}_i,\lambda}+C_{13}(\underline{k}_i)b^+_{\underline{k}_i,\lambda}+C_{14}(\underline{k}_i)b_{-\underline{k}_i,\lambda} \quad ,$$

$$\alpha^{(1)}_{-\underline{k}_i,\lambda} = C_{21}(\underline{k}_i)a^+_{\underline{k}_i,\lambda}+C_{22}(\underline{k}_i)a_{-\underline{k}_i,\lambda}+C_{23}(\underline{k}_i)b^+_{\underline{k}_i,\lambda}+C_{24}(\underline{k}_i)b_{-\underline{k}_i,\lambda} \quad , \tag{4-189}$$

and two corresponding relations for $\alpha^{(2)+}_{\underline{k}_i,\lambda}$ and $\alpha^{(2)}_{-\underline{k}_i,\lambda}$, respectively. The coefficients are interrelated $C_{21} = C^*_{12}$, $C_{22} = C^*_{11}$, $C_{23} = C^*_{14}$, $C_{24} = C^*_{13}$. Because these new operators must behave as creation and annihilation operators, they must diagonalize the Hamilton operator

$$H = \sum_{\substack{\underline{k}_i \\ \lambda=1,2}} \Big[\hbar\omega^{(1)}(\underline{k}_i)(\alpha^{(1)+}_{\underline{k}_i,\lambda}\alpha^{(1)}_{\underline{k}_i,\lambda}+1/2)+\hbar\omega^{(2)}(\underline{k}_i)(\alpha^{(2)+}_{\underline{k}_i,\lambda}\alpha^{(2)}_{\underline{k}_i,\lambda}+1/2)\Big] \tag{4-190}$$

and fulfill the commutation rules

$$\Big[\alpha^{(g)}_{\underline{k}_i},\alpha^{(h)+}_{\underline{k}_j}\Big] = \delta_{\underline{k}_i,\underline{k}_j}\cdot\delta_{gh} \quad , \tag{4-191}$$

$g,h = 1,2$ and $i,j = 1,\dots,N$. The index λ can be omitted because only photons and phonons with the same polarization direction will couple. Conditions allowing the determination of the coefficients $C_{\ell,m}$ ($\ell,m = 1,2,3,4$) in (4-189) are derived from (4-191), e.g.

$$\Big[\alpha^{(1)}_{\underline{k}_i},\alpha^{(1)+}_{\underline{k}_i}\Big] = C_{11}C_{22} - C_{12}C_{21} + C_{13}C_{24} - C_{14}C_{23} =$$

$$= |C_{11}|^2-|C_{12}|^2+|C_{13}|^2-|C_{14}|^2 = 1 \quad . \tag{4-192}$$

From (4-190) and (4-191) we further derive the important commutation rules

$$(1/\hbar)\left[\alpha_{\underline{k}_i}^{(1)+}, H\right] = -\omega^{(1)}(\underline{k}_i)\alpha_{\underline{k}_i}^{(1)+} \quad , \tag{4-193}$$

$$(1/\hbar)\left[\alpha_{\underline{k}_i}^{(1)}, H\right] = \omega^{(1)}(\underline{k}_i)\alpha_{\underline{k}_i}^{(1)} \quad , \tag{4-194}$$

and corresponding relations for $\alpha_{\underline{k}_i}^{(2)+}$ and $\alpha_{\underline{k}_i}^{(2)}$. The frequencies $\omega^{(1)}(\underline{k}_i)$ and $\omega^{(2)}(\underline{k}_i)$ are completely determined by (4-191) through (4-194). We show below that they are identical to those calculated on the basis of the classic wave model for polaritons, see (3-31). The operators defined by (4-189) therefore describe creation and annihilation of the quantized mixed electromagnetic-mechanical waves. As pointed out in 1.1, we also refer to these quanta as 'polaritons'. The index (1) refers to the phonon-like dispersion branch and (2) to the photon-like one.

When determining the coefficients $C_{\ell,m}(\underline{k}_i)$ and the polariton frequencies $\omega^{(g)}(\underline{k}_i)$, we can omit the index i on the wave-vector because only states with identical $\underline{k}_i$ will couple, see (4-189) and (4-191). Substituting (4-189) into (4-193) leads to

$$\left[-\omega_P C_{11}(\underline{k}) - iDC_{13}(\underline{k}) - iDC_{14}(\underline{k})\right] a_{\underline{k}}^{+} + \left[+\omega_P C_{12}(\underline{k}) - iDC_{13}(\underline{k}) - iDC_{14}(\underline{k})\right] a_{-\underline{k}} +$$

$$+\left[\ iDC_{11}(\underline{k}) - iDC_{12}(\underline{k}) - \omega_L C_{13}(\underline{k})\right] b_{\underline{k}}^{+} + \left[-iDC_{11}(\underline{k}) + iDC_{12}(\underline{k}) + \omega_L C_{14}(\underline{k})\right] b_{-\underline{k}} =$$

$$= -\omega^{(1)}(\underline{k})\left[C_{11}(\underline{k}) a_{\underline{k}}^{+} + C_{12}(\underline{k}) a_{-\underline{k}} + C_{13}(\underline{k}) b_{\underline{k}}^{+} + C_{14}(\underline{k}) b_{-\underline{k}}\right] \quad . \tag{4-195}$$

From the condition that the factors of $a_{\underline{k}}^{+}$, $a_{-\underline{k}}$, $b_{\underline{k}}^{+}$ and $b_{-\underline{k}}$ on both sides of (4-195) have to be identical, we derive an eigenvalue equation of the following form

$$\begin{pmatrix} -\omega_P+\omega^{(1)}(\underline{k}) & 0 & -iD & -iD \\ 0 & \omega_P+\omega^{(1)}(\underline{k}) & -iD & -iD \\ iD & -iD & -\omega_L+\omega^{(1)}(\underline{k}) & 0 \\ -iD & iD & 0 & \omega_L+\omega^{(1)}(\underline{k}) \end{pmatrix} \begin{pmatrix} C_{11} \\ C_{12} \\ C_{13} \\ C_{14} \end{pmatrix} = 0 \quad . \tag{4-196}$$

The determinant of the coefficient matrix in (4-196) has to vanish for nontrivial solutions. ω_P is obviously identical with the frequency ω_L of the longitudinal phonon. This requires

$$\left[\omega^{(1)}(\underline{k})\right]^4-\left[\omega^{(1)}(\underline{k})\right]^2(\omega_P^2+\omega_L^2)+\omega_P^2\omega_L^2-4\omega_P\omega_L D^2 = 0 \qquad (4\text{-}197)$$

or, taking (4-164), (4-185) and (4-187) into account,

$$\left[\omega^{(1)}(\underline{k})\right]^4-\left[\omega^{(1)}(\underline{k})\right]^2(\varepsilon_o\omega_T^2+c^2k^2)+\omega_T^2c^2k^2 = 0$$

or

$$\left(c^2k^2/[\omega^{(1)}]^2\right) = \frac{\varepsilon_o\omega_T^2-[\omega^{(1)}]^2}{\omega_T^2-[\omega^{(1)}]^2} \quad . \qquad (4\text{-}198)$$

Eq.(4-198), however, is identical to (3-17) if we there substitute $\varepsilon_\infty = 1$. Hence, both the particle and the wave pictures yield the same dispersion relation for polaritons. The polariton frequency does not depend on the wave-vector direction in cubic crystals. Hence in the following we only consider the magnitude of $\underline{k}$.

Finally, we determine the coefficients $C_{\ell,m}$, which obviously represent the magnitudes of the respective contributions of phonons and photons to the polariton.

The factor which is left open by (4-196) is determined by (4-192) if e.g. we assert that C_{13} shall be a real quantity. Using Cramer's rule, we get

$$C_{11}(k) = -C_{22}(k) = \frac{-i}{2\sqrt{\omega_P}}\left(\omega_P+\omega^{(1)}(\underline{k})\right)\left(-\omega_L^2+[\omega^{(1)}(\underline{k})]^2\right)A^{-1} \quad ,$$

$$C_{12}(k) = -C_{21}(k) = \frac{-i}{2\sqrt{\omega_P}}\left(-\omega_P+\omega^{(1)}(\underline{k})\right)\left(-\omega_L^2+[\omega^{(1)}(\underline{k})]^2\right)A^{-1} \quad ,$$

$$C_{13}(k) = C_{24}(k) = -D\sqrt{\omega_P}(\omega_L+\omega^{(1)}(k))A^{-1} \quad , \qquad (4\text{-}199)$$

$$C_{14}(k) = C_{23}(k) = D\sqrt{\omega_P}(-\omega_L+\omega^{(1)}(k))A^{-1}$$

where

$$A(\omega^{(1)}(k)) = \sqrt{\omega^{(1)}(k)\{(-\omega_L^2+[\omega^{(1)}(k)]^2)^2+4D^2\omega_L\omega_P\}} \quad . \qquad (4\text{-}200)$$

Corresponding relations are derived for the coefficients C_{i1}, C_{i2}, C_{i3} and C_{i4} ($i = 3,4$) of the operators $\alpha^{(2)+}$ and $\alpha^{(2)}$ referring to the photon-like branch.

There is no principal theoretical difficulty in extending the quantum field theory for polaritons to include electronic polarization ($\varepsilon_\infty = 1$) or arbitrary crystals. However, the amount of calculation involved is enormously increased. For this reason, we omit any further discussions of this kind.

5. Some Special Topics Relative to Polaritons

5.1 STIMULATED RAMAN SCATTERING BY POLARITONS

The difference between spontaneous and stimulated Raman scattering may be characterized by the following statement. The stimulated Raman effect is a scattering process associated with optical feedback amplifying the scattered radiation whereas with spontaneous Raman scattering no feedback occurs.

We compare the simplest one dimensional classic models in some detail so as to demonstrate the physical background.

1) Spontaneous Raman scattering by phonons: an incident laser wave $E(\omega_L) = E_L \exp(-i\omega_L t)$ enters a crystal and is modulated by a phonon wave $Q(\omega_p) = Q\exp(-i\omega_p t)$. Taking into account Placzek's expansion of the polarizability $\alpha(Q)$ with respect to the vibrational normal coordinate Q, the radiating dipole moments are

$$M = \left[\alpha_o + (\partial\alpha/\partial Q)_o \mathrm{Re}Q(\omega_p) + (1/2)(\partial^2\alpha/\partial Q^2)_o \mathrm{Re}Q(\omega_p) + \ldots\right] \mathrm{Re}E(\omega_L) \quad . \qquad (5\text{-}1)$$

The first term describes Rayleigh scattering whereas the linear term corresponds to the Stokes and Anti-Stokes radiation at the frequencies $\omega_S = \omega_L - \omega_p$ and $\omega_A = \omega_L + \omega_p$, respectively. Terms of higher order describe higher-order Raman effects at frequencies $\omega = \omega_L \pm n \cdot \omega_p$ $(n = 2,3,\ldots)$. The energy density of the medium contains a mechanical part due to the lattice waves and terms representing the potential energy of dipoles $M_i = \alpha_o E_i$ induced by the electric fields E_i. The indices i = L, S and A stand for Laser, Stokes and Anti-Stokes, respectively. These dipoles build up the macroscopic polarizations, hence $P_i = \chi_o E_i$. χ_o denotes the linear susceptibility of the medium. The energy density thus can be introduced as

$$\Phi = (1/2)Q^*(\omega_p)\omega_p^2 Q(\omega_p) - (1/2)E^*(\omega_L)\chi_o E(\omega_L) -$$

$$-(1/2)E^*(\omega_S)\chi_o E(\omega_S) - (1/2)E^*(\omega_A)\chi_o E(\omega_A) + c.c. \quad . \tag{5-2}$$

Laser, Stokes and Anti-Stokes radiation in the medium can be described by wave equations of type

$$\Delta E - (1/c^2)\ddot{E} = (4\pi/c^2)\ddot{P} \tag{5-3}$$

if the material is assumed to be isotropic, i.e. div $\underline{E} = 0$. The field variables therefore are treated as scalars. The polarizations P_i of the three waves are derived from (5-2) as $-\partial\Phi/\partial E^*(\omega_L)$, $-\partial\Phi/\partial E^*(\omega_S)$, $-\partial\Phi/\partial E^*(\omega_A)$, and the equation of motion as

$$\ddot{Q} + \Gamma\dot{Q} = -\partial\Phi/\partial Q^*(\omega_p) \quad , \tag{5-4}$$

see 3.4. Damping of the phonons is taken into account. The situation after a spontaneous Raman scattering process in the medium is described by the following set of wave equations:

$$\Delta E(\omega_L) - (\varepsilon_L/c^2)\ddot{E}(\omega_L) = 0 \quad , \tag{5-5a}$$

$$\Delta E(\omega_S) - (\varepsilon_S/c^2)\ddot{E}(\omega_S) = 0 \quad , \tag{5-5b}$$

$$\Delta E(\omega_A) - (\varepsilon_A/c^2)\ddot{E}(\omega_A) = 0 \quad , \tag{5-5c}$$

$$\ddot{Q}(\omega) + \Gamma\dot{Q}(\omega) + \omega_p^2 Q(\omega_p) = 0 \quad . \tag{5-5d}$$

This can easily be verified by taking $\varepsilon E = E + 4\pi P$. The equations (5-5) are completely decoupled, which implies no interaction between the waves after the scattering process, in other words, principle of superimposition known from linear optics is valid here.

2) Stimulated Raman scattering by phonons: Primary scattering processes take place in the same way as before. However, the scattered radiation is now rather strong, being caused by a powerful exciting (laser) source, and the nonlinear terms of the polarization have also to be taken into account:

$$P = \left[\chi_o + (\partial\chi/\partial Q)_o \mathrm{Re}Q(\omega_p) + \ldots\right] \mathrm{Re}E(\omega_L) \quad . \tag{5-6}$$

The energy density (5-2) has to be generalized by four nonlinear terms

$$\Phi = (1/2)Q^*(\omega_p)\omega_p^2 Q(\omega_p) - (1/2)E^*(\omega_L)\chi_o E(\omega_L) -$$

$$-(1/2)E^*(\omega_S)\chi_o E(\omega_S) - (1/2)E^*(\omega_A)\chi_o E(\omega_A) -$$

$$-E^*(\omega_S)(\partial\chi/\partial Q)Q^*(\omega_p)E(\omega_L) - E^*(\omega_A)(\partial\chi/\partial Q)Q(\omega_p)E(\omega_L) + c.c. \qquad (5\text{-}7)$$

The derivates $\partial\chi/\partial Q$ and later $\partial\chi/\partial E$ have always to be determined in the equilibrium as $(\partial\chi/\partial Q)_o$ and $(\partial\chi/\partial E)_o$. Laser, Stokes and Anti-Stokes waves in the medium are again described by wave equations of type (5-3), but now with polarizations derived from (5-7). The linear parts of these may be included in the dielectric constant whereas the nonlinear parts remain on the right-hand sides of the equations. The system corresponding to (5-5a to d) therefore becomes

$$\Delta E(\omega_L) - \frac{\varepsilon_L}{c^2}\ddot{E}(\omega_L) = \frac{4\pi}{c^2}\frac{\partial^2}{\partial t^2}\frac{\partial\chi}{\partial Q}(Q(\omega_p)E(\omega_S) + Q^*(\omega_p)E(\omega_A)) \quad , \qquad (5\text{-}8a)$$

$$\Delta E(\omega_S) - \frac{\varepsilon_S}{c^2}\ddot{E}(\omega_S) = \frac{4\pi}{c^2}\frac{\partial^2}{\partial t^2}\frac{\partial\chi}{\partial Q}Q^*(\omega_p)E(\omega_L) \quad , \qquad (5\text{-}8b)$$

$$\Delta E(\omega_A) - \frac{\varepsilon_A}{c^2}\ddot{E}(\omega_A) = \frac{4\pi}{c^2}\frac{\partial^2}{\partial t^2}\frac{\partial\chi}{\partial Q}Q(\omega_p)E(\omega_L) \quad , \qquad (5\text{-}8c)$$

$$\ddot{Q}(\omega) + \Gamma\dot{Q}(\omega) + \omega_p^2 Q(\omega) = (\partial\chi/\partial Q)(E(\omega_L)E^*(\omega_S) + E(\omega_A)E^*(\omega_L)) \quad . \qquad (5\text{-}8d)$$

In contrast to (5-5a to d) all these equations are coupled. From the equation of motion (5-8d) there follows for resonance

$$Q(\omega_p) = \frac{\partial\chi}{\partial Q}\frac{1}{-i\omega\Gamma}(E(\omega_L)E^*(\omega_S) + E(\omega_A)E^*(\omega_L)) \quad . \qquad (5\text{-}9)$$

So the lattice wave is driven by electromagnetic fields at the frequency $\omega_L - \omega_S = \omega_A - \omega_L = \omega_p$. This causes a stronger modulation of the incident laser wave and consequently amplified Stokes and Anti-Stokes radiation. As a result the lattice wave is again more excited, and so on. The mechanism corresponds to optical feedback (at the resonance frequency ω_p) and is limited only by the damping Γ.

The macroscopic polarization becomes

$$P = (1/2)\chi_o\left[E(\omega_L)+E^*(\omega_L)\right]+(1/2)(\partial\chi/\partial Q)\ \frac{1}{-i\omega_p\Gamma}\left[E^2(\omega_L)E^*(\omega_S)+\right.$$

$$\left.+E_L^2E(\omega_A)+E^2(\omega_L)E^*(\omega_A)+E_L^2E(\omega_S)+c.c.\right] \quad , \qquad (5\text{-}10)$$

when taking (5-9) into account. E_L denotes the amplitude of $E(\omega_L)$. The factor

$$\frac{\partial\chi}{\partial Q} \cdot \frac{1}{-i\omega_p\Gamma} = \chi_{NL} \qquad (5\text{-}11)$$

is referred to as the nonlinear susceptibility of the medium for the mechanism in question. χ_{NL} obviously becomes purely imaginary for full resonance. According to (5-10), the polarizations at the frequencies ω_S and ω_A are directly proportional to the square of the laser field. For weak laser intensities, as in the case of spontaneous Raman scattering, the corresponding terms may be neglected because χ_{NL} is small. For strong laser fields, however, the terms can dominate the linear term and cause amplification of the Stokes and Anti-Stokes waves. Equations (5-8a to d) may be simplified by the assumption that the fraction of energy transferred from the laser wave to the Stokes and Anti-Stokes waves is small, i.e. the amplitude E_L is treated as constant so that the laser wave equation (5-8a) can be omitted.

The set (5-8b through d) can be solved for an 'almost' one-dimensional case by considering the experimental situation as follows.

All waves are polarized in z direction. The incident laser beam propagates in x direction. Stokes and Anti-Stokes radiation in the xy plane are considered for near-forward directions a few degrees away from the x direction so that an amplification of the waves is approximately proportional to the x component of the propagation length in the medium. The (small) y components of the wave-vectors $\underline{k}_S$ and $\underline{k}_A$ are neglected. A corresponding ansatz for the laser, Stokes, and Anti-Stokes waves becomes

$$E(\underline{r},\omega_L) = E_L\exp\{ik_{Lx}^o\cdot x-i\omega_Lt\} \qquad (5\text{-}12)$$

$$E(\underline{r},\omega_S) = E_S\exp\{i\left[(k_{Sx}^o+ik_{Sx}')x+k_{Sy}^o\cdot y\right]-i\omega_St\} \qquad (5\text{-}13)$$

$$E(\underline{r},\omega_A) = E_A\exp\{i\left[(k_{Ax}^o+ik_{Ax}')x+k_{Ay}^o\cdot y\right]-i\omega_At+i\phi(x)\} \quad . \qquad (5\text{-}14)$$

The x components of the wave-vectors of the two scattered waves thus have been introduced as complex. The imaginary parts determine the exponents of the factors describing the amplification or attenuation of E_S and E_A. A x dependent phase factor has been added to the Anti-Stokes wave; this factor is proportional to the travel length in the medium and the exponent can be written $\phi(x) = \Delta k \cdot x$. Δk may be interpreted as a 'wave-vector mismatch' of k^o_{Ax}. Eqs.(5-13) and (5-14) are rewritten to become

$$E(\underline{r},\omega_S) = E_S \exp[-k'_{Sx}\cdot x] \, \exp[i(k^o_{Sx}\cdot x + k^o_{Sy}\cdot y) - i\omega_S t] \quad , \tag{5-15}$$

and

$$E(\underline{r},\omega_A) = E_A \exp[-k'_{Ax}\cdot x] \exp[i\Delta k\cdot x] \exp[i(k^o_{Ax}\cdot x + k^o_{Ay}\cdot y) - i\omega_A t] \quad . \tag{5-16}$$

If we substitute (5-12), (5-15) and (5-16) into (5-8b) and (5-8c) while taking (5-9) into account, the wave-vector mismatch turns out to be

$$\Delta k = 2k^o_{Lx} - k^o_{Sx} - k^o_{Ax} \quad . \tag{5-17}$$

This result easily is derived when the amplitudes are considered to depend only weakly on x so that the second derivates $\Delta \equiv \partial^2/\partial x^2$ can be neglected. The two coupled differential equations then reduce to an algebraic system for the amplitudes E_S and E^*_A or E^*_S and E_A. When making the approximation $k'_{Sx} \approx k'_{Ax} \equiv \gamma$ for the 'gain factor', and $k^o_{Ax}/\omega^2_A \approx k^o_{Sx}/\omega^2_S$, the zeros of the determinant of this system require

$$\gamma = i\left\{\frac{\Delta k}{2} \pm \left[\left(\frac{\Delta k}{2}\right)^2 - i\,\frac{2\pi\omega^2_S}{k^o_{Sx}c^2}\left(\frac{\partial\chi}{\partial Q}\right)^2 \frac{1}{\omega_p\Gamma}\,|E_L|^2\Delta k\right]^{-1/2}\right\} \tag{5-18}$$

For $\Delta k >> (2\pi\omega^2_S/k^o_{Sx}c^2)\,\mathrm{Im}\chi_{NL}|E_L|^2$ the ratio $|E_S/E^*_A|^2$ is found to be $>> 1$, which implies that only the Stokes wave is amplified. The Anti-Stokes wave will be attenuated.

For $\Delta k \approx (2\pi\omega^2_S/k^o_{Sx}c^2)\,\mathrm{Im}\chi_{NL}|E_L|^2$, on the other hand, $|E_S/E^*_A|^2 \approx 4$. Both scattered waves are strongly amplified.

Finally, for $\Delta k = 0$ we get $\gamma = k'_{Ax} = k'_{Sx} = 0$. This result shows that for perfect phase matching no stimulated amplification takes place at all. The wave-vector mismatch Δk has necessarily to be introduced in order to explain the appearance of stimulated Anti-

Stokes radiation.

Bloembergen has shown /168/ that Δk should be of the order of $\Delta k \approx 2(2\pi\omega_S^2/k_{Sx}^o c^2)\,\mathrm{Im}\chi_{NL}|E_L|^2$ for maximum amplification of the Anti-Stokes wave. If Δk becomes larger, only the Stokes wave will again be strongly generated (see above). Detailed discussions are given in /167, 168, 344/. A wave-vector diagram corresponding to our ansatz is shown in Fig.44. Only the momentum mismatch of the Anti-Stokes wave in x direction has been considered. The left part of the figure corresponds to a Stokes process

$$\underline{k}_L-\underline{k}_p-\underline{k}_S = 0 \tag{5-19}$$

whereas the right part illustrates a 'mismatched' Anti-Stokes process of the form

$$\underline{k}_L+\underline{k}_p-\underline{k}_A-\Delta\underline{k} = 0 \quad . \tag{5-20}$$

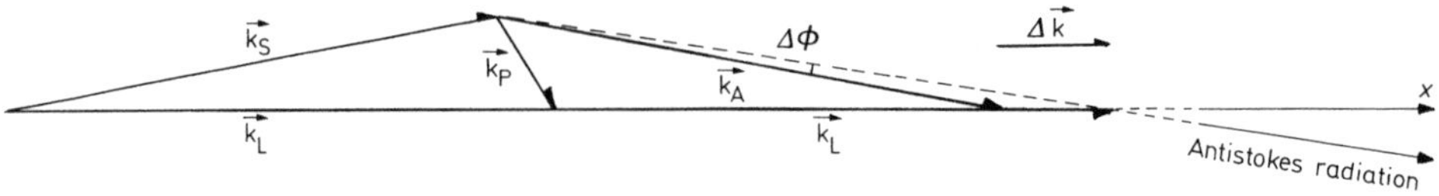

Fig.44 Wave-vector diagram illustrating the generation of stimulated Anti-Stokes radiation in the direction $\underline{k}_A+\Delta\underline{k} = 2\underline{k}_L-\underline{k}_S$, see text.

Stimulated Stokes scattering processes can take place for arbitrary directions of $\underline{k}_S$. If the laser frequency ω_L lies in the visible, the phonons created have wave-vectors with magnitudes in the region 10^4 to $\sim 10^5$ cm^{-1} and directions determined by (5-19). If polariton dispersion effects are neglected, all these phonons have the same frequency. A stimulated Anti-Stokes process, however, can only take place for a large number of excited phonons fitting the wave-vector relation (5-20). Hence stimulated Anti-Stokes radiation is emitted only in certain directions: $\underline{k}_A+\Delta\underline{k}$, as illustrated in Fig.44. By adding (5-19) and (5-20), we obtain

$$\Delta\underline{k} = 2\underline{k}_L-\underline{k}_S-\underline{k}_A \tag{5-21}$$

which is equivalent to (5-17).

In practice, the wave-vector mismatch is very small, corresponding to $\Delta\phi \approx 0.5^{\circ}$ in Fig.44. The intensity dip expected for $\Delta \underline{k} \equiv 0$ has not so far been experimentally observed. Stimulated Anti-Stokes radiation thus appears in near-forward directions conically around the exciting laser beam for $\Delta \underline{k} \approx 0$. This result has been experimentally verified, see /169-172/.

A stimulated Raman spectrum in general /173/ consists of a strong fundamental mode and some of its harmonics, which do not show frequency shifts due to anharmonicities like two-phonon processes in second-order spontaneous Raman scattering, see 1.5. The fundamental mode is usually that with the largest scattering cross-section and possibly small damping. For increasing exciting laser intensities the Stokes wave caused by this phonon must first become strong enough to induce optical feedback. The Stokes radiation finally becomes extremely strong and may itself act as a pumping beam and induce stimulated radiation at the frequency $\omega_L - 2\omega_p$. The first harmonic mode created in this way thus does not show any frequency shifts due to anharmonicities because only a first-order scattering process excited by the stimulated Stokes wave is primarily involved. Fig.45 is a schematic diagram demonstrating the generation of the harmonics due to this mechanism. The intense or amplified light beams are indicated by thick arrows and the weak or attenuated beams by thin arrows.

The theory outlined so far has been concerned only with stimulated Raman scattering by long-wavelength phonons without polariton dispersion. The situation for polaritons has been discussed in, e.g. /163, 167 and 174/. A frequency-dependent electric field $E(\omega)$ at (approximately) the phonon frequency has to be considered in addition. The macroscopic susceptibility χ then depends on the quasi-normal coordinate $Q(\omega)$ as well as on $E(\omega)$ and Placzek's expansion in the linear approximation becomes

$$\chi(Q,E) \approx \chi_o + (\partial\chi/\partial Q)Q(\omega) + (\partial\chi/\partial E)E(\omega) \quad . \tag{5-22}$$

$\partial\chi/\partial Q$ denotes an atomic displacement coefficient and $\partial\chi/\partial E$ an electro-optic coefficient as before, see 4.12. The energy density (5-7) now has to be generalized by terms describing the coupling between the polar phonon mode and the electric field $E(\omega)$. Furthermore terms

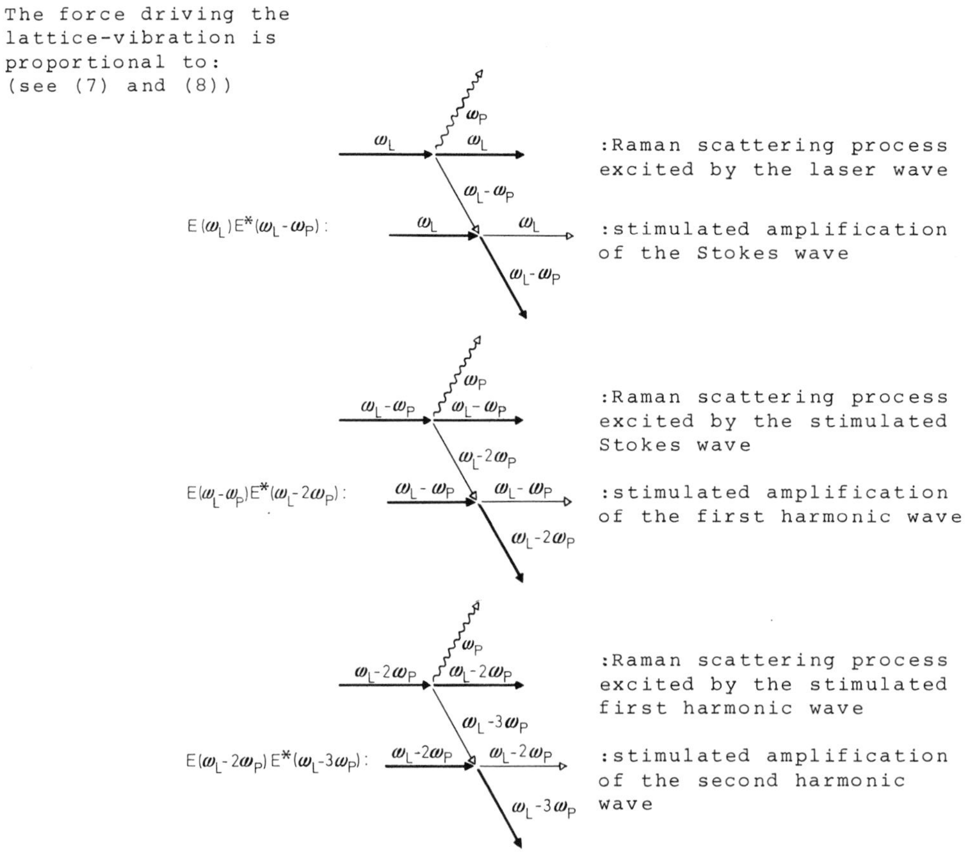
The force driving the lattice-vibration is proportional to: (see (7) and (8))
E(ωL)E*(ωL-ωP):
:Raman scattering process excited by the laser wave
:stimulated amplification of the Stokes wave
E(ωL-ωP)E*(ωL-2ωP):
:Raman scattering process excited by the stimulated Stokes wave
:stimulated amplification of the first harmonic wave
E(ωL-2ωP)E*(ωL-3ωP):
:Raman scattering process excited by the stimulated first harmonic wave
:stimulated amplification of the second harmonic wave

Fig.45 Diagrams demonstrating a mechanism for the generation of harmonics in a stimulated Raman spectrum. The intense or amplified light beams are indicated by thick arrows.

must be introduced to describe the potential energy of dipoles oscillating at the polariton frequency in the field $E(\omega)$, and four nonlinear terms originating from the electro-optic coefficient term in (5-22). The energy density becomes explicitly

$$\begin{aligned}\Phi = &-(1/2)Q^*(\omega)B^{11}Q(\omega)-Q^*(\omega)B^{12}E(\omega)-(1/2)E^*(\omega)B^{22}E(\omega)-\\ &-(1/2)E^*(\omega_L)B^{22}E(\omega_L)-(1/2)E^*(\omega_S)B^{22}E(\omega_S)-(1/2)E^*(\omega_A)B^{22}E(\omega_A)-\\ &-E^*(\omega_S)(\partial\chi/\partial Q)Q^*(\omega)E(\omega_L)-E^*(\omega_A)(\partial\chi/\partial Q)Q(\omega)E(\omega_L)-\\ &-E^*(\omega_S)(\partial\chi/\partial E)E^*(\omega)E(\omega_L)-E^*(\omega_A)(\partial\chi/\partial E)E(\omega)E(\omega_L)+c.c. \qquad (5\text{-}23)\end{aligned}$$

We have reintroduced the coefficients $B^{11} = -\omega_p^2$ and $B^{22} = \chi_o$ in order to show the connection with the theory outlined in 4.1. (Note that $\chi_o = (\varepsilon_\infty - 1)/4\pi$, which can be seen directly from $\varepsilon_\infty \cdot E = E + 4\pi P$.) The polarizations and the equation of motion are derived in the same way as before, so that the generalized set of equations for stimulated polaritons corresponding to (5-8a through d) becomes

$$\Delta E(\omega_L) - \frac{\varepsilon_L}{c^2}\ddot{E}(\omega_L) = \frac{4\pi}{c^2}\frac{\partial^2}{\partial t^2}\left[\frac{\partial\chi}{\partial Q}(Q(\omega)E(\omega_S)+Q^*(\omega)E(\omega_A)) + \right.$$

$$\left. + \frac{\partial\chi}{\partial E}(E(\omega)E(\omega_S)+E^*(\omega)E(\omega_A))\right] , \qquad (5\text{-}24a)$$

$$\Delta E(\omega_S) - \frac{\varepsilon_S}{c^2}\ddot{E}(\omega_S) = \frac{4\pi}{c^2}\frac{\partial^2}{\partial t^2}\left[\frac{\partial\chi}{\partial Q}Q^*(\omega)E(\omega_L) + \frac{\partial\chi}{\partial E}E^*(\omega)E(\omega_L)\right] , \qquad (5\text{-}24b)$$

$$\Delta E(\omega_A) - \frac{\varepsilon_A}{c^2}\ddot{E}(\omega_A) = \frac{4\pi}{c^2}\frac{\partial^2}{\partial t^2}\left[\frac{\partial\chi}{\partial Q}Q(\omega)E(\omega_L) + \frac{\partial\chi}{\partial E}E(\omega)E(\omega_L)\right] , \qquad (5\text{-}24c)$$

$$\Delta E(\omega) - \frac{\varepsilon_\infty}{c^2}\ddot{E}(\omega) = \frac{4\pi}{c^2}\frac{\partial^2}{\partial t^2}\left[B^{12}Q(\omega) + \frac{\partial\chi}{\partial E}E(\omega_L)E^*(\omega_S) + \frac{\partial\chi}{\partial E}E^*(\omega_L)E(\omega_A)\right] , \qquad (5\text{-}24d)$$

$$P(\omega_L) = B^{22}E(\omega_L) + \frac{\partial\chi}{\partial Q}(Q(\omega)E(\omega_S)+Q^*(\omega)E(\omega_A)) +$$

$$+ \frac{\partial\chi}{\partial E}(E(\omega)E(\omega_S)+E^*(\omega)E(\omega_A)) , \qquad (5\text{-}24e)$$

$$P(\omega_S) = B^{22}E(\omega_S) + \frac{\partial\chi}{\partial Q}Q^*(\omega)E(\omega_L) + \frac{\partial\chi}{\partial E}E^*(\omega)E(\omega_L) , \qquad (5\text{-}24f)$$

$$P(\omega_A) = B^{22}E(\omega_A) + \frac{\partial\chi}{\partial Q}Q(\omega)E(\omega_L) + \frac{\partial\chi}{\partial E}E(\omega)E(\omega_L) , \qquad (5\text{-}24g)$$

$$P(\omega) = B^{12}Q(\omega) + B^{22}E(\omega) + \frac{\partial\chi}{\partial E}(E^*(\omega_S)E(\omega_L)+E(\omega_A)E^*(\omega_L)) , \qquad (5\text{-}24h)$$

$$\ddot{Q}(\omega) + \Gamma\dot{Q}(\omega) = B^{11}Q(\omega) + B^{12}E(\omega) + \frac{\partial\chi}{\partial Q}(E^*(\omega_S)E(\omega_L)+E(\omega_A)E^*(\omega_L)) . \qquad (5\text{-}24i)$$

Eqs.(5-24h) and (5-24i) are Huang's equations and include damping for stimulated scattering, see (3-9), (3-10), and (4-83), (4-84). Eqs.(5-24a through i) describe the stimulated Raman effect for polaritons in isotropic crystals. The approximation curl curl $\underline{E} = -\Delta\underline{E}$ can be made only for exactly transverse $\underline{E}$ fields. This condition is fulfilled in cubic crystals and also holds for ordinary polaritons and the principal direction in uniaxial crystals. Extraordinary polaritons, however, are in general of mixed type LO + TO, so that the equations corresponding to (5-24a through d) must be derived from

$$\text{curl curl } \underline{E} + (1/c^2)\ddot{\underline{E}} = -(4\pi/c^2)\ddot{\underline{P}} \quad . \tag{5-25}$$

The set (5-24) can be solved by assuming that the laser field amplitude remains constant as before and neglecting the Anti-Stokes wave. Eq.(5-24) then reduces to

$$\Delta E(\omega_S) - \frac{\varepsilon_S}{c^2}\ddot{E}(\omega_S) = \frac{4\pi}{c^2}\frac{\partial^2}{\partial t^2}\left[\frac{\partial\chi}{\partial Q}Q^*(\omega)E(\omega_L) + \frac{\partial\chi}{\partial E}E^*(\omega)E(\omega_L)\right] \quad , \tag{5-26a}$$

$$\Delta E(\omega) - \frac{\varepsilon_\infty}{c^2}\ddot{E}(\omega) = \frac{4\pi}{c^2}\frac{\partial^2}{\partial t^2}\left[B^{12}Q(\omega) + \frac{\partial\chi}{\partial E}E(\omega_L)E^*(\omega_S)\right] \quad , \tag{5-26b}$$

$$P(\omega_S) = B^{22}E(\omega_S) + \frac{\partial\chi}{\partial Q}Q^*(\omega)E(\omega_L) + \frac{\partial\chi}{\partial E}E^*(\omega)E(\omega_L) \quad , \tag{5-26c}$$

$$P(\omega) = B^{12}Q(\omega) + B^{22}E(\omega) + \frac{\partial\chi}{\partial E}E^*(\omega_S)E(\omega_L) \quad , \tag{5-26d}$$

$$\ddot{Q}(\omega) + \Gamma\dot{Q}(\omega) = B^{11}Q(\omega) + B^{12}E(\omega) + \frac{\partial\chi}{\partial Q}E^*(\omega_S)E(\omega_L) \quad . \tag{5-26e}$$

This system is linear because E_L = const. The linearity allows us to introduce a plane-wave ansatz for $E^*(\omega_S)$, $P^*(\omega_S)$, $E(\omega)$, $P(\omega)$, and $Q(\omega)$. These quantities thus become proportional to $\exp(-ik_Sx+i\omega_St)$ and $\exp(ikx-i\omega t)$, respectively, and the determinant of the system (5-26) is

$$\begin{vmatrix}
-(k_S^*)^2 + \frac{\varepsilon_S\omega_S^2}{c^2} & 0 & \frac{4\pi\omega_S^2}{c^2}\frac{\partial\chi}{\partial E}E_L^* & 0 & \frac{4\pi\omega_S^2}{c^2}\frac{\partial\chi}{\partial Q}E_L^* \\
\frac{4\pi\omega^2}{c^2}\frac{\partial\chi}{\partial E}E_L & 0 & -(k)^2 + \frac{\varepsilon_\infty\omega^2}{c^2} & 0 & \frac{4\pi\omega^2}{c^2}B^{12} \\
-B^{22} & 1 & -\frac{\partial\chi}{\partial E}E_L^* & 0 & -\frac{\partial\chi}{\partial Q}E_L^* \\
-\frac{\partial\chi}{\partial E}E_L & 0 & -B^{22} & 1 & -B^{12} \\
-\frac{\partial\chi}{\partial Q}E_L & 0 & -B^{12} & 0 & -(\omega^2+i\omega\Gamma+B^{11})
\end{vmatrix} \quad .$$

The zeros of this determinant which can be derived without any difficulty lead to the generalized dispersion relation of polaritons for stimulated Stokes scattering in isotropic materials

$$\left[(c^2k^2/\omega^2)-\varepsilon_1\right]\left[(c^2k_S^{*2}/\omega_S^2)-\varepsilon_2\right]-\varepsilon_3^2 = 0 \quad . \tag{5-27}$$

The three dielectric constants introduced for abbreviation are

$$\varepsilon_1 = \varepsilon_\infty - 4\pi (B^{12})^2 (\omega^2 + i\omega\Gamma + B^{11})^{-1} \quad ,$$

$$\varepsilon_2 = \varepsilon_S - 4\pi (\partial\chi/\partial Q)^2 E_L^2 (\omega^2 + i\omega\Gamma + B^{11})^{-1} \quad ,$$

and

$$\varepsilon_3 = 4\pi (\partial\chi/\partial E) E_L - 4\pi (\partial\chi/\partial Q) B^{12} E_L (\omega^2 + i\omega\Gamma + B^{11})^{-1} \quad .$$

$\varepsilon_S \equiv \varepsilon(\omega_S)$ in the second relation can be set $= \varepsilon_\infty$ because the Stokes frequency is large compared with the polariton frequency. The interpretation of these dielectric constants becomes more obvious if we look at the relations

$$D(\omega) = \varepsilon_1 E(\omega) + \varepsilon_3 E^*(\omega_S) \tag{5-28a}$$

and

$$D(\omega_S) = \varepsilon_2 E(\omega_S) + \varepsilon_3 E^*(\omega) \quad . \tag{5-28b}$$

They can easily be derived from

$$D(\omega) = E(\omega) + 4\pi P(\omega)$$

and

$$D(\omega_S) = E(\omega_S) + 4\pi P(\omega_S)$$

by taking into account (5-26c,d,e) and a plane-wave ansatz. For small exciting laser fields where no stimulated scattering takes place, i.e. $E_L \to 0$, (5-27) reduces to

$$\left[(c^2k^2/\omega^2) - \varepsilon_\infty + 4\pi (B^{12})^2 (\omega^2 + i\omega\Gamma + B^{11})^{-1}\right]\left[(c^2 k_S^{*2}/\omega^2) - \varepsilon_\infty\right] = 0 \quad . \tag{5-29}$$

The zeros of the first bracket are equivalent to (3-31), (4-12) or (4-75). They describe the polariton dispersion when no nonlinear effects are involved. The second bracket describes the 'dispersion' of light, which is linear if ε_∞ is regarded as constant.

The real parts of the solutions of (5-27) determine the wave-vectors for the modified polaritons in nonlinear materials, while the wave-vectors of the modified Stokes waves are obtained in the same way, taking momentum and energy conservation into account: $\underline{k}_L = \underline{k}_S + \underline{k}$ and $\omega_L = \omega_S + \omega$. The imaginary parts of the wave-vectors as before describe the amplification or attenuation of the waves.

Eq.(5-27) has been discussed quantitatively by Rath /174/ for GaP. He neglected mechanical damping: $\Gamma = 0$. The evaluation shows that close to the lattice resonance ω_p the polariton frequency is hardly changed. This fact has frequently been verified because spontaneous and stimulated scattering by polar phonons are observed at the same frequency within the experimental error. Differences, however, occur in regions where the energy of the polaritons contains a large electromagnetic contribution, i.e. for $\omega \gtrsim \omega_{LO}$ and $\omega \lesssim \omega_{TO}$.

An extensive review article on theoretical and experimental work concerning stimulated Raman scattering up to 1967 has been published by Bloembergen /175/. Stimulated scattering by polaritons was first observed by Tannenwald and Weinberg on quartz in 1967 /288/, see also a note by Scott /289/. The dispersion of the nonlinear susceptibility in the polariton region of GaP was investigated by Faust et al. in 1968 /176/. In 1969 Kurtz and Giordmaine observed the stimulated Raman effect on the polariton mode associated with the strong A_1(TO) phonon at $\sim$ 628 cm^{-1} in $LiNbO_3$ /177/. The frequency shift observed in this material was found to be in agreement with the dispersion curve calculated for spontaneous scattering. Gelbwachs et al. /179/ first reported tunable stimulated emission of Stokes and Anti-Stokes waves excited by the polariton mode associated with the A_1(TO) phonon at 248 cm^{-1} in $LiNbO_3$. These authors were still using an external resonator, which prevented them from observing the highly coherent infrared emission in the direction of the stimulated polariton wave. Only a few months later Yarborough and coworkers succeeded in generating stimulated emission from the same polariton mode without an external resonator /178/; they observed the stimulated Stokes wave as well as the corresponding idler wave in the IR region. Variation of the scattering angle allowed tuning of the radiation in the visible region from 42 to 200 cm^{-1} and in the infrared region from 238 to 50 μ. The conversion of the pump beam to the tunable frequencies was found to be more than 50 %. The experiment has been improved so that Johnson et al. in 1971 reported peak powers outside the crystal of approximately 3 W at the idler wavelength of 200 μ /180/. Linewidth measurements showed a bandwidth of less than 0.5 cm^{-1} over the tuning range of 66 to 200 μ.

From the beginning these experiments were of great interest for the construction of tunable Raman lasers, especially in the IR region, and experimental studies to find other suitable materials have been going on. Thus, stimulated emission from the polariton mode associated with the strong A(TO) phonon at 793 cm^{-1} in $LiIO_3$ was reported almost simultaneously in 1971 by Schrötter /181/ and Amzallag et al. /182/. They observed frequency shifts down to 759 and 766cm^{-1}, respectively. The absorption of the idler wave in this material, however, was rather large. An additional result of special interest reported by Schrötter was the observation of stimulated emission from a second harmonic Stokes component of the polariton mode at 1506 cm^{-1}, see Fig.46.

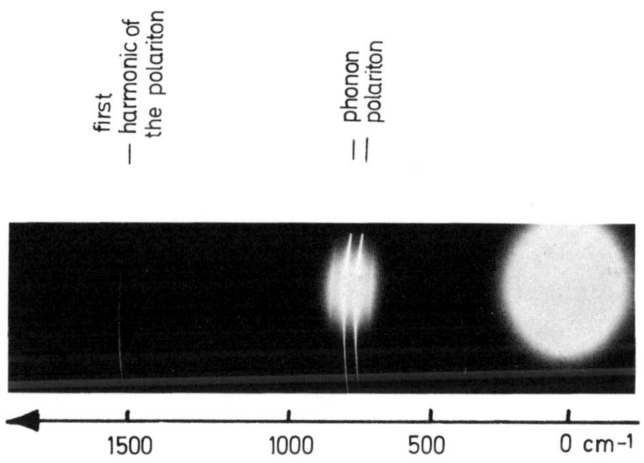

Fig.46 Stimulated emission from a polariton associated with the A(TO) phonon at 793 cm^{-1} and its first harmonic in $LiIO_3$, from /181/.

Another technique for the construction of a polariton laser in the IR region was proposed and first demonstrated by De Martini /183-185/. Although these experiments do not directly involve stimulated scattering, the subject is rather closely related to this nonlinear effect. The author used two exciting laser beams in the visible region at frequencies ω_1 and ω_2. The energy difference of the quanta was chosen to be identical with the energy of a polariton $\hbar\omega = \hbar\omega_1 - \hbar\omega_2$ in a nonlinear material (here GaP). If the two laser beams enter the sample so as to form an angle such that the momentum-matching condition $\underline{k} = \underline{k}_1 - \underline{k}_2$ for polariton is fulfilled, strong infrared radiation is observed at the difference frequency $\omega_1 - \omega_2$ in the direction of the polariton wave-vector.

A similar experiment has recently been reported by Hwang and Solin /186/. They report mixing of exciting argon-laser radiation at the frequency ω_1 in the visible, and carbon-dioxide laser radiation at ω_2 in the IR region with $RbClO_3$ as nonlinear medium. The polariton mode is directly driven by the incident IR radiation and is not due only to nonlinear terms such as $(\partial\chi/\partial E)E^*(\omega_S)E(\omega_L)$ in (5-26e), which would require a giant pulse incident laser with $\omega_L = \omega_1$. Both the Stokes and Anti-Stokes components of the hot polaritons have been observed at $\omega_1-\omega_2$ and $\omega_1+\omega_2$, respectively. Although modes with small scattering intensities in the spontaneous spectra do not generally cause stimulated emission, the experiment of Hwang and Solin shows that, at least in the polariton region of polar phonons, the generation of 'quasi-stimulated' radiation from such modes is possible.

'Coherent Anti-Stokes Raman spectroscopy' /306/ is another example of wave mixing caused by nonlinearities in matter. A strong light beam from, for instance, a Nd:YAG laser at the frequency ω_L is partly used to pump a dye laser. Radiation from the latter is tuned to coincide with the Stokes frequency ω_S of a Raman signal from a nonlinear medium. The remaining pump beam at ω_L and the dye laser beam are weakly focused and crossed inside the material so that the phase-matching condition $2\underline{k}_L-\underline{k}_S \approx \underline{k}_A$, (5-21), for stimulated Anti-Stokes radiation is fulfilled, see also Fig.44 and related text. As a result radiation is generated at the Anti-Stokes frequency $\omega_A = 2\omega_L-\omega_S$ and may easily be analyzed with a spectrometer because it leaves the sample at an angle which is obviously almost twice the crossing angle $\sphericalangle(\underline{k}_L, \underline{k}_S)$.

We finally draw attention to experiments concerning stimulated scattering excited by picosecond pulses, reported by Lauberau et al. /187/, Karmenyan et al. /188/ and Herrmann et al. /192/. The materials used were GaP and $LiIO_3$. Because the exciting laser pulses were so short, direct observation of the lifetime of polaritons was possible.

Additional Literature

Faust, W.L., Henry, C.H.: Mixing of visible and near-resonance infrared light in GaP /213/.

Hwang, D.M., Solin, S.A.: Raman scattering from tunable hot phonons and polaritons /238/.

Strizhevskii, V.L., Yashkir, Yu.N.: Peculiarities of Raman scattering by polaritons and polar optical phonons in anisotropic crystals /241/.

Biraud-Laval, S.: Dispersion des polaritons ordinaires dans le quartz /283/.

De Martini, F.: Nonlinear optical investigation of polaritons in solids /308/.

Hollis, R.L.,Ryan,I.F.,Scott,I.F.: Spin-Flip-Plus-Phonon Raman scattering: A new second-order scattering process /349/.

5.2 POLARITON-PLASMON COUPLING

The vibrational energy quanta of an electron plasma in the conduction band of a material are known as plasmons. Plasmons are purely longitudinal and have a resonance frequency ω_P given by

$$\omega_P = (4\pi\nu e^2/\varepsilon_\infty m^*)^{1/2} \quad , \tag{5-30}$$

where ν denotes the electron density in cm^{-3}, e the elementary charge, ε_∞ the dielectric constant at high frequencies as before, and m* the effective mass of the electrons (or holes) in the conducting band. A coupling of plasmons with phonons is expected only for purely longitudinal polar modes because transverse phonons and plasmons are linearly independent. Strong interaction takes place when the density of carriers ν determines a plasma frequency ω_P approaching that of a polar LO phonon $\omega_P \approx \omega_L$. Varga /189/ has shown that the total dielectric function in the long-wavelength limit is then built up additively by contributions originating from the valence electrons, the polar lattice vibrations, and the conducting electrons

$$\varepsilon(\omega) = \varepsilon_\infty + \omega_T^2 \left[(\varepsilon_0 - \varepsilon_\infty)/(\omega_T^2 - \omega^2)\right] - \omega_P^2 \varepsilon_\infty/\omega^2 \quad . \tag{5-31}$$

Damping has been neglected. Eq.(5-31) is identical with (3-27) except for the additional Drude term. The (longitudinal) eigenfrequencies of the system are determined by the zeros of the dielectric function $\varepsilon(\omega)$. Eq.(5-30) and (5-31) thus describe the dispersion ω versus carrier concentration ν for the coupled system (plasmon-LO phonon)

$$(4\pi e^2/m^*)(\nu/\omega^2) = \varepsilon_\infty \left[(\omega_L^2 - \omega^2)/(\omega_T^2 - \omega^2)\right] \quad . \tag{5-32}$$

This relation is formally identical with the dispersion relation of polaritons in the diatomic cubic case, see 3.3 (k^2 is replaced by ν and c^2 by $4\pi e^2/m^*$). The curves ω versus $\sqrt{\nu}$ are similar in shape to the polariton dispersion curves, see /189/. The theory has so far been experimentally verified by Mooradian and Wright in GaAs /19/.

The dispersion relation for the modified transverse modes in the presence of an electron plasma is derived from (5-31) in the usual way by substituting $\varepsilon(\omega) = c^2k^2/\omega^2$

$$(c^2k^2/\omega^2) = \varepsilon_\infty\left[(\omega_L^2-\omega^2)/(\omega_T^2-\omega^2)-\omega_P^2/\omega^2\right] \quad . \tag{5-33}$$

In the center of the 1BZ the frequencies are

$$\omega^2 = (1/2)(\omega_P^2+\omega_L^2) \pm \left[(1/4)(\omega_P^2+\omega_L^2)-\omega_P^2\omega_T^2\right]^{1/2} \quad . \tag{5-34}$$

The 'photon-like' polariton branch ends at a frequency slightly higher than ω_L and the lowest-frequency branch does not reach $\omega = 0$ for $k \to 0$. No photon propagation is possible for frequencies below the plasma frequency in a conducting material, see Fig.47a and b. For vanishing carrier densities ν, $\omega_P = 0$ and (5-34) again describes the situation in insulators.

The polariton dispersion curves are changed in shape because a longitudinal mode appears in addition at approximately the plasma eigenfrequency and because the frequency of the polar LO phonon is somewhat raised due to coupling with the plasmons. (The frequencies in question depend on the carrier concentration ν, see (5-30) and (5-34).) Polaritons, however, are not directly coupled to the (longitudinal) plasmons as they are purely transverse. The normal coordinates are linearly independent.

In order to achieve direct coupling with polariton modes in cubic materials, a partly transverse character has to be induced in the plasmons by external parameters, e.g. a magnetic field $\underline{B}$. If $\underline{v}$ denotes the velocity of the electrons, the plasmon wave-vector is parallel to $\underline{v}$. As can be seen from the Lorentz-force term $e(\underline{v}\times\underline{B})$, the plasmons will in fact no longer be longitudinal when a magnetic field is applied perpendicular to $\underline{v}$. The coupled state photon-phonon-plasmon has been called plasmariton. The phonon-like polariton branch splits into two components as in Fig.47c. Experiments to

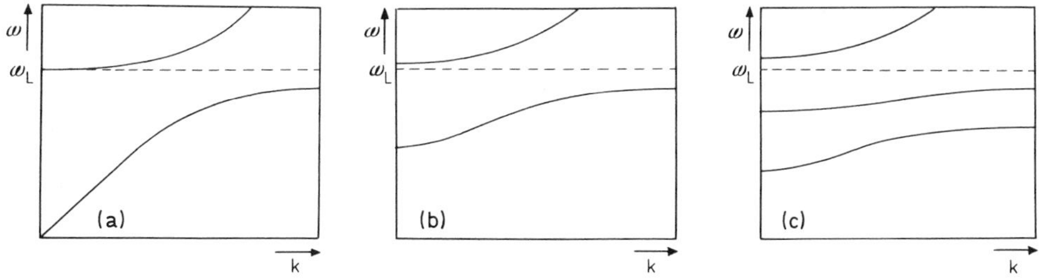

Fig.47 a) Polariton dispersion branches in the cubic diatomic lattice of an insulator.
b) Polariton dispersion in the presence of an electron plasma.
c) Plasmariton dispersion branches when a magnetic field is applied, see text, from /190/.

show the existence of the coupled polariton-plasmon state described above have been carried out by Patel and Slusher on GaAs /190/. Good agreement was found with the upper of the two phonon-like branches whereas the other one at $\sim$ 50 cm^{-1} lower wave numbers has not yet been observed. The term 'plasmariton', has also been used for the excitations described by the dispersion curves in Fig.47b, where no direct coupling takes place between polaritons and plasmons, in a paper by Shah et al. /191/. The experiments reported concern ordinary polaritons in uniaxial CdS and correspond to plasmaritons in the limiting case $\underline{B} = 0$.

Finally, we note that the observation of plasmaritons as sketched in Fig.47c should be possible in noncubic material with free carriers without application of a magnetic field. According to 4.7, extraordinary polaritons are in general of mixed type LO + TO for arbitrary directions of the wave-vectors $\underline{k}$, the LO and TO components changing magnitude with the wave-vector direction. Thus, couplings of different strengths should be observed when the angle between $\underline{k}$ and the optic axis is varied. No experiments of this kind have so far been published.

5.3 ON THE OBSERVATION OF BULK POLARITONS BY TM REFLECTION

Two recent papers by Falge, Otto, and Sohler /193/ and Nitsch, Falge, and Claus /194/ present a new experimental technique that allows the observation of bulk polaritons by infrared reflection spectroscopy.

The dispersion relation for polaritons is determined by the dielectric functions of a crystal, see 3.3 and 4.2. As is well-known, the dielectric functions essentially determine the structure of the IR reflection spectra /195, 196/. We therefore derive the latter as a function of the wave-vector $\underline{k}_{BP}$ of bulk polaritons. As a result it turns out that in suitable experiments the turning points of the edges of the reflection bands depend on this wave-vector in the same way as the bulk polariton frequency.

The principle of the experimental setup is sketched in Fig.48. An incident electromagnetic wave ($\underline{E}_i$) is partly reflected ($\underline{E}_{rfl}$) and partly refracted ($\underline{E}_{rfr}$) by the surface of a crystalline medium. The optic axis (z) stands normal to the surface. $\underline{E}_{rfr}$ corresponds to the extraordinary ray in the material. The surrounding (upper) medium with the refractive index n_1 is assumed to be isotropic. All light waves are (electrically) polarized in the plane of the figure, i.e. the magnetic polarization is perpendicular to the optic axis and the figure plane. These conditions have become known as 'transverse magnetic polarization' (TM polarization). With the ansatz

$$\underline{E}_i = E_i \begin{pmatrix} \cos\alpha \\ 0 \\ \sin\alpha \end{pmatrix} \exp\left[i(k_x, -k_z)\begin{pmatrix} x \\ z \end{pmatrix} - i\omega t\right] ,$$

and

$$\underline{E}_{rfl} = E_{rfl} \begin{pmatrix} -\cos\alpha \\ 0 \\ \sin\alpha \end{pmatrix} \exp\left[i(k_x, k_z)\begin{pmatrix} x \\ z \end{pmatrix} - i\omega t\right]$$

where $k_x = (\omega/c)n_1 \sin\alpha$ and $k_z = (\omega/c)n_1 \cos\alpha$, the reflectivity can be derived /195,196/ as

$$R_{TM}(\alpha,\omega) = \left|\frac{\underline{E}_{rfl}}{\underline{E}_i}\right|^2 = \left|\frac{\sqrt{\varepsilon_x}\sqrt{\varepsilon_z}(\omega/c)\cos\alpha - n_1\sqrt{\varepsilon_z(\omega/c)^2 - k_x^2}}{\sqrt{\varepsilon_x}\sqrt{\varepsilon_z}(\omega/c)\cos\alpha + n_1\sqrt{\varepsilon_z(\omega/c)^2 - k_x^2}}\right|^2 . \quad (5\text{-}35)$$

Fig.48 Incident ($\underline{E}_i$), reflected ($\underline{E}_{rfl}$) and refracted ($\underline{E}_{rfr}$) ray when TM reflection is being measured in order to determine the dispersion of bulk polaritons, see text, from /194, 197/.

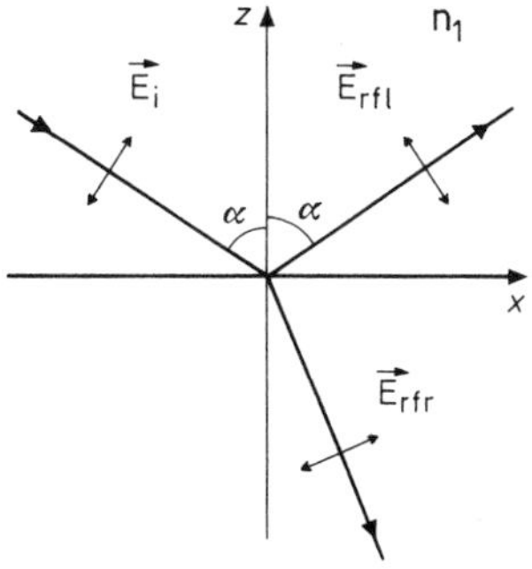

Continuity in the surface z = 0 has to be required for (a) the x components of the wave-vector and the electric field and (b) the y component of the magnetic field. Maxwell's equations have also to be taken into account.

The real z component of the wave-vector of the refracted ray can be determined from the generalized Fresnel equation (4-20):

$$k_z^{(rfr)} = (1/\varepsilon_z)\left[(\varepsilon_z(\omega^2/c^2)-k_x^2)\varepsilon_x\varepsilon_z\right]^{1/2} \quad . \tag{5-36}$$

For extraordinary bulk polaritons with wave-vectors $\underline{k}_{BP}$ propagating perpendicular to the optic axis (x direction of Fig.48), we have according to (4-54)

$$c^2k_{BP}^2/\omega^2 = \varepsilon_z \tag{5-37}$$

where $\varepsilon_z \equiv \varepsilon_{\parallel}$. Combining (5-35) and (5-37), the reflectivity becomes

$$R_{TM}(\alpha,\omega) = \left|\frac{\sqrt{\varepsilon_x}\ k_{BP}\ \cos\alpha - n_1\sqrt{k_{BP}^2-k_x^2}}{\sqrt{\varepsilon_x}\ k_{BP}\ \cos\alpha + n_1\sqrt{k_{BP}^2-k_x^2}}\right|^2 \quad . \tag{5-38}$$

The angle α is defined as indicated in Fig.48. Total reflection takes place if one, but not both, of the terms appearing in both the numerator and the denominator of (5-38) is imaginary. This happens for $\varepsilon_z > 0$ when either ε_x or $(k_{BP}^2 - k_x^2)$ is negative, and for $\varepsilon_z < 0$ when simultaneously $\varepsilon_x < 0$. In all other cases reduced reflectivity takes place with a refracted ray $\underline{E}_{rfr}$. According to (5-36) and (5-37), the real wave-vector component in direction z becomes

$$k_z^{(rfr)} = (1/\varepsilon_z)\sqrt{(k_{BP}^2 - k_x^2)\varepsilon_x\varepsilon_z} \tag{5-39}$$

for the refracted ray. This wave couples with the optical phonons propagating in direction z.

We describe experiments on $K_3Cu(CN)_4$ /194/. The polariton dispersion curves of this material in the high-frequency region originating from C ≡ N stretching vibrations are reproduced in Fig.49a. Areas with total reflection determined by the conditions given above have been hatched in the diagram. According to (5-39) the limiting curves separating areas of total and reduced reflection are identical with the dispersion curves of bulk polaritons propagating parallel to the surface. Their lattice displacements are parallel to the z direction. This is easily verified by considering (5-39) for $k_z^{(rfr)} = 0$. Recording IR- or ATR-reflection spectra /198, 199/ for different fixed angles $\alpha = \alpha_0$ gives reflectivity scans corresponding to the traces indicated in the diagram for $\alpha = 24^\circ$ and $\alpha = 32^\circ$.

Damping has so far been neglected in this experiment. When damping is taken into account, the edges of the total reflection areas become smoothed. In real experiments the turning points of the edges of the reflection bands will correspond to the bulk polariton-dispersion curves /197/. Two scans for $K_3Cu(CN)_4$ corresponding to $\alpha = 24^\circ$ and $\alpha = 32^\circ$ are reproduced in Figs.49b and 49c. For better orientation, one typical turning point has been indicated by a small square in Fig.49a and c. The two extremely narrow areas of total reflection in the $\alpha = 24^\circ$ trace could not be observed, see Fig.49a.

The method described has proved to be an excellent complement and sometimes superior to the common Raman scattering technique, see /194/.

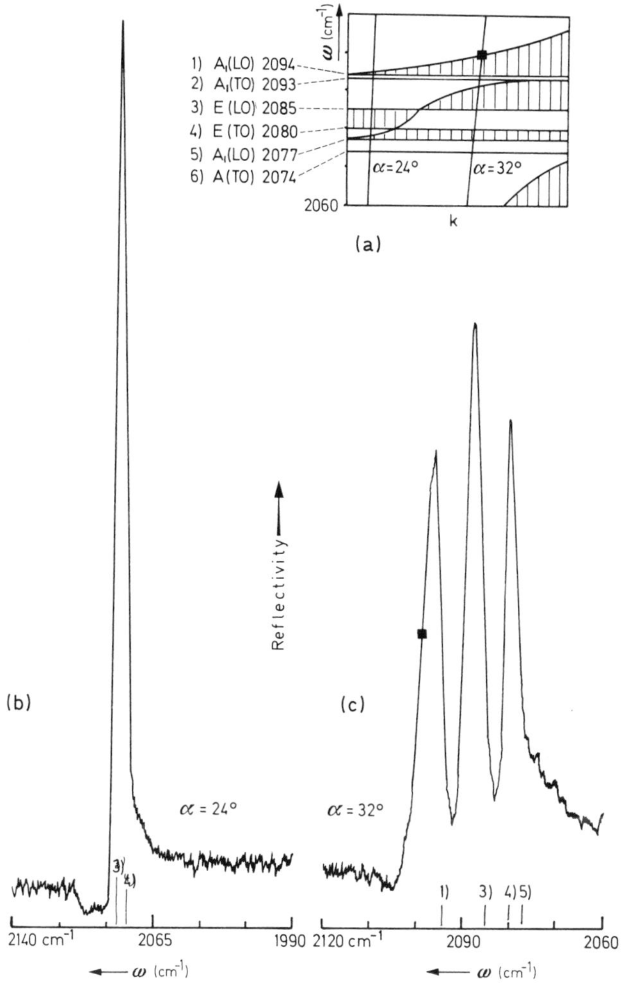

Fig.49 a) ω-k diagram of the high-frequency extraordinary polariton region in $K_3Cu(CN)_4$ for wave-vector directions perpendicular to the optic axis. Areas with total reflection are hatched.

b) IR-reflection scan corresponding to $\alpha = 24^o$ as indicated in a).

c) ATR-reflection spectrum (n_1: KRS 5) corresponding to $\alpha = 32^o$ as indicated in a), from /194, 197/. The angles α are defined as shown in Fig.48.

5.4 SURFACE POLARITONS

The dispersion effects of polaritons localized at surfaces of insulating media have become of great interest within the last few years. We refer to an extensive article on the subject by Borstel et al. /253/ for a complete list of references and detailed discussions of recent developments. We deal here only with the principle of the phenomenon in order to show the connection with the properties derived for bulk polaritons earlier in this article.

For an isotropic dielectric medium the two 'rotational' Maxwell's equations determine the wave equation

$$\text{curl curl } \underline{E} = -(\varepsilon(\omega)/c^2)\ddot{\underline{E}} \quad , \tag{5-40}$$

see (5-3). By introducing an ansatz of the form

$$\underline{E} = \underline{E}(z)\exp(ik_x x - i\omega t) \tag{5-41}$$

describing electromagnetic plane waves in direction x along the surface z = 0, we obtain the following three relations for the dependence of the amplitude $\underline{E}(z)$ on the distance from the surface (z = 0)

$$\partial^2 E_x(z)/\partial z^2 = ik_x(\partial E_z(z)/\partial z) - (\varepsilon(\omega)\omega^2/c^2)E_x(z) \tag{5-42}$$

$$\partial^2 E_y(z)/\partial z^2 = (k_x^2 - \varepsilon(\omega)\omega^2/c^2)E_y(z) \quad , \tag{5-43}$$

and

$$ik_x(\partial E_x(z)/\partial z) = -(k_x^2 - \varepsilon(\omega)\omega^2/c^2)E_z(z) \quad . \tag{5-44}$$

Taking into account that $E_x(z) = E_y(z)$ because of the isotropy, we make an ansatz for the amplitudes outside the medium (z > 0), where the refractive index is assumed to be n = 1 (vacuum)

$$E_x(z) = E_x \exp\left[-(k_x^2 - \omega^2/c^2)^{1/2}\right]z \quad , \tag{5-45}$$

and

$$E_z(z) = ik_x(k_x^2 - \omega^2/c^2)^{1/2} \cdot E_z \exp\left[-(k_x^2 - \omega^2/c^2)^{1/2}\right]z \quad . \tag{5-46}$$

Inside the medium (z < 0) correspondingly we state

$$E_x(z) = E_x \exp\left[+(k_x^2-\varepsilon\omega^2/c^2)^{1/2}\right] z \quad , \tag{5-47}$$

and

$$E_z(z) = ik_x(k_x^2-\varepsilon\omega^2/c^2)^{1/2}\cdot E_z \exp\left[+(k_x^2-\varepsilon\omega^2/c^2)^{1/2}\right] z \quad . \tag{5-48}$$

Note that in the medium $n^2 = \varepsilon(\omega)$. The quantities $(k_x^2-\omega^2/c^2)^{1/2}=k_{zV}$ and $(k_x^2-\varepsilon(\omega)\cdot\omega^2/c^2)^{1/2} = k_{zM}$ are obviously the complex wave-vector components in direction z describing the decay of the amplitudes in vacuum (index = V) and in the medium (index = M), respectively, see Fig.50. The signs are negative for z > 0 and positive for z < 0, see (5-45) to (5-48). At the surface (z = 0) the $E_x(0)$ and the $D_z(0)$ components both have to change continuously. The first condition leads to the triviality $E_x(0) = E_x(0)$ whereas the second one requires

$$iE_z\, k_x(k_x^2-\omega^2/c^2)^{-1/2} = -i\varepsilon(\omega)k_x(k_x^2-\varepsilon(\omega)\omega^2/c^2)^{-1/2}\cdot E_z \quad . \tag{5-49}$$

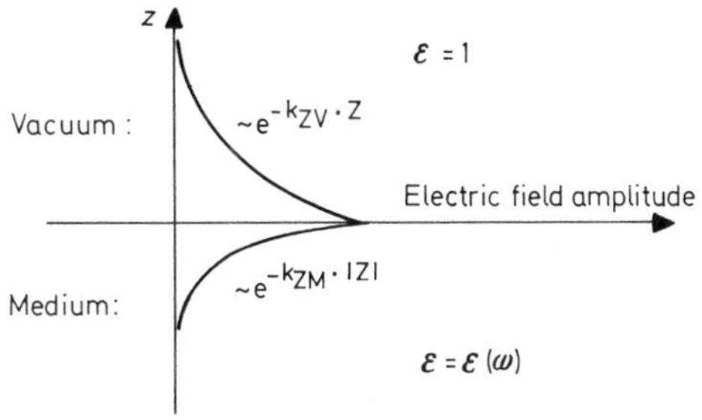

Fig.50: Pattern showing the decay of the electric field amplitude as a function of the distance z from the surface z = 0, from /290/.

From (5-49) we derive the dispersion relation of surface polaritons

$$(c^2k_x^2/\omega^2) = \frac{\varepsilon(\omega)}{\varepsilon(\omega)+1} \quad . \tag{5-50}$$

The difference from the corresponding result for bulk polaritons is obviously the appearance of the denominator $\varepsilon(\omega)+1$ on the right-hand side. The condition for the existence of surface phonons directly follows from (5-50) for $k_x \to \infty$, as $\varepsilon(\omega) = -1$, a result familiar from the literature /257/. Note that the denominator $\varepsilon(\omega)+1$ in (5-50) originates from the introduction of the continuity conditions at the surface. The z components of the wave-vectors are derived from (5-50) by taking into account the definitions $k_{zV} = (k_x^2-\omega^2/c^2)^{1/2}$ and $k_{zM} = (k_x^2-\varepsilon\omega^2/c^2)^{1/2}$

$$k_{zV} = (\omega/c)\left[-(\varepsilon(\omega)+1)^{-1}\right]^{1/2} \quad , \tag{5-51}$$

$$k_{zM} = (\omega/c)\left[-\varepsilon^2(\omega)(\varepsilon(\omega)+1^{-1}\right]^{1/2} \quad . \tag{5-52}$$

From these relations it follows that surface polaritons exist only for dielectric functions $\varepsilon(\omega) < -1$, because k_{zV} and k_{zM} are required to be real. Fig.51 shows a surface polariton-dispersion curve determined by (5-50) for a simple diatomic crystals. The branch lies in the interval between the frequencies of the transverse (ω_T) and longitudinal (ω_L) optic phonons. This energy region is not covered by bulk polaritons. Investigations of surface polaritons thus represent a perfect complement to studies of bulk polaritons /253, 290/. Taking into account the dielectric function given by (3-27), the maximum frequency ω_s of the surface polariton can be derived from $\varepsilon(\omega) = -1$ as

$$\omega_s^2 = \left[(\varepsilon_o+1)/(\varepsilon_\infty+1)\right]\omega_T^2 \quad . \tag{5-53}$$

This frequency can easily be changed by varying the refractive index of the surrounding medium. When instead of vacuum ($\varepsilon = 1$) the space over the surface is a medium with the dielectric constant $\varepsilon = \varepsilon'$, the limiting frequency ω_s becomes

$$\omega_s^2 = \left[(\varepsilon_o+\varepsilon')/(\varepsilon_\infty+\varepsilon')\right]\omega_T^2 \quad . \tag{5-54}$$

A detailed discussion has been presented by Otto /260/. For appropriate experimental methods the reader is again referred to /253/.

A first observation of surface polaritons by attenuated total reflection was reported by Marschall and Fischer in 1972 /198/. The material was GaP, the same as was used by Henry and Hopfield /7/ seven years earlier for the first observation of bulk polaritons. Falge and Otto /258/ clarified the situation in uniaxial polyatomic crystals, the material investigated being α quartz. The three different geometries studied by these authors are sketched in Fig.52.

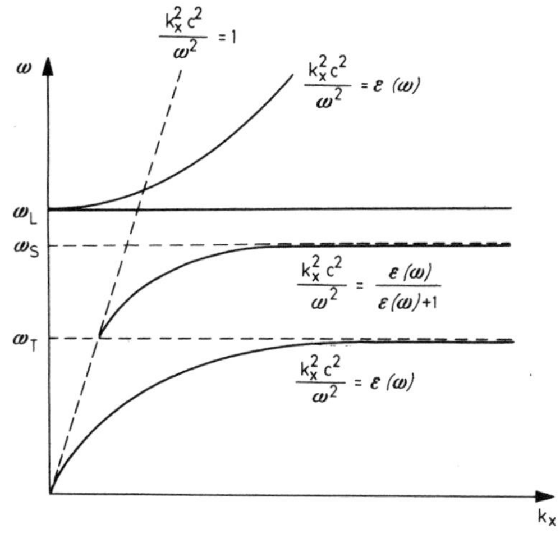

Fig.51 Surface polariton dispersion curve as determined by (5-50) and bulk polaiton-dispersion curves for a diatomic cubic material.

When we examine (5-45) through (5-48), it becomes obvious that the main properties of surface polaritons in uniaxial crystals can be derived by taking into account only two components of the electric field vector, one parallel to the wave-vector direction and the other normal to the surface. In Fig.52a both components are oriented perpendicular to the optic axis (z). This implies that only the dielectric function $\varepsilon_\perp(\omega)$ is involved in the dispersion relation

$$(c^2k^2/\omega^2) = \varepsilon_\perp(\omega)/(\varepsilon_\perp(\omega)+1) \quad , \tag{5-55}$$

see /253/. Eq.(5-55) is obviously identical to (5-50) for isotropic media. In analogy to the denomination of bulk polaritons in uniaxial crystals, these excitations have been named ordinary surface polaritons. They exist for $\varepsilon_\perp(\omega) < -1$ and $k > \omega/c$, the same conditions as for cubic crystals.

In the other geometries illustrated in Fig.52b and c there is one component of the electric field vector parallel to the optic axis and another perpendicular to it. The dispersion relations derived are

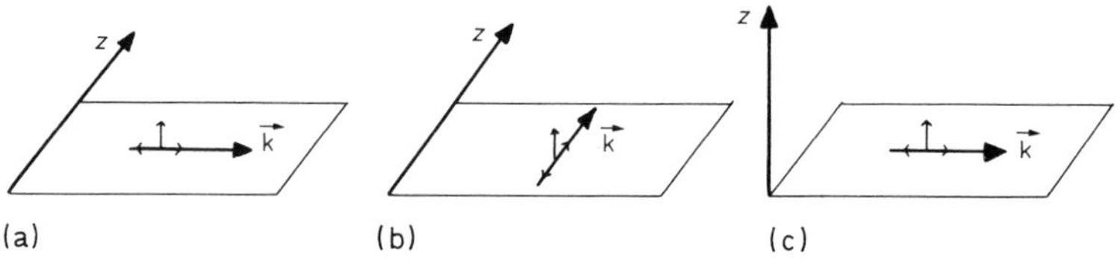

Fig.52 Three geometric situations for the observation of surface polaritons in uniaxial crystals: a) $\underline{k}$ lying in an optically isotropic plane and propagating perpendicularly to the optic axis (z): ordinary surface polaritons are observed. b) and c) correspond to extraordinary surface polaritons.

$$(c^2k^2/\omega^2) = (\varepsilon_{\parallel}(\omega)\varepsilon_{\perp}(\omega)-\varepsilon_{\perp}(\omega))/(\varepsilon_{\parallel}(\omega)\varepsilon_{\perp}(\omega)-1) \tag{5-56}$$

for Fig.52b (with the conditions $\varepsilon_{\parallel}(\omega) < 0$, $k > \omega/c$), and

$$(c^2k^2/\omega^2) = (\varepsilon_{\parallel}(\omega)\varepsilon_{\perp}(\omega)-\varepsilon_{\parallel}(\omega))/(\varepsilon_{\parallel}(\omega)\varepsilon_{\perp}(\omega)-1) \tag{5-57}$$

for Fig.52c (with the conditions $\varepsilon_{\perp}(\omega) < 0$, $k > \omega/c$).

Eqs.(5-56) and (5-57) describe special cases of <u>extraordinary surface polaritons</u> in the same way as (4-54) and (4-55) do for bulk polaritons. We restrict our discussion to these limiting cases since they have hitherto proved to be of the greatest interest for experiments. The general case for arbitrary wave-vector directions has been discussed in /259, 261, 262/.

Because (5-56) and (5-57) are symmetrical with respect to the indices $\perp$ and $\parallel$, we consider only Fig.52b in more detail. The results can easily be transferred to the other geometry. Real and virtual (excitation) surface polaritons are distinguished:

a) In frequency regions where $\varepsilon_{\parallel}(\omega) < 0$ and $\varepsilon_{\perp}(\omega) < 0$ the dispersion relation determines a maximum frequency ω_s derived from

$$\varepsilon_{\parallel}(\omega)\varepsilon_{\perp}(\omega) = 1 \tag{5-58}$$

for $k \to \infty$. The corresponding modes are referred to as <u>real surface polaritons.</u>

b) Virtual (excitation) surface polaritons exist in frequency regions where $\varepsilon_{\parallel}(\omega) < 0$ and $\varepsilon_{\perp}(\omega) > 1$. The extraordinary surface polariton branch starts at the frequency ω' where a bulk polariton branch described by $(c^2k^2/\omega^2) = \varepsilon_{\perp}(\omega)$ crosses the light line $(c^2k^2/\omega^2) = 1$ and ends at the crossing point of the bulk polariton curve with a longitudinal optical phonon branch at $\omega = \omega_{\parallel L}$. It can be shown that the transverse and longitudinal parts of the extraordinary surface polariton at this latter crossing point disintegrate into a transverse bulk polariton and a longitudinal optical phonon propagating parallel to the surface.

In regions where $\varepsilon_{\parallel}(\omega) > 0$ or $\varepsilon_{\parallel}(\omega) < 0$ and $0 < \varepsilon_{\perp}(\omega) < 1$, there are no extraordinary surface polaritons. Fig.53 qualitatively shows the dielectric functions $\varepsilon = \varepsilon_{\parallel}(\omega)$ and $\varepsilon = \varepsilon_{\perp}(\omega)$ of α quartz for $300 < \omega < 600\ \mathrm{cm}^{-1}$. In order to illustrate the conditions given above, regions with real and virtual surface polaritons (for the geometry of Fig.52b) are indicated by hatched areas in the lower part of the figure. Forbidden regions are indicated above. Corresponding experimental results have been published by Falge and Otto /258/.

Additional Literature

Schopper, H.: Zur Optik dünner doppelbrechender und dichroitischer Schichten /244/.

Flournoy, P.A., Schaffers, W.J.: Attenuated total reflection spectra from surfaces of anisotropic absorbing films /245/.

Mosteller, L.P. jr., Wooten, F.: Optical properties and reflectance of uniaxial crystals /246/.

Decius, J.C., Frech, R., Brüesch, P.: Infrared reflection from longitudinal modes in anisotropic crystals /247/.

Sohler, W.: Extension of Wolter's multilayer reflection and transmission formulae for orthorhombic absorbing media /248/.

Szivessy, G.: Polarisation /249/.

Wolter, H.: Schlieren-, Phasenkontrast- und Lichtschrittverfahren/250/.

Bell, E.E.: Optical constants and their measurement /251/.

Koch, E.E.,Otto,A.,Kliewer,K.L.: Reflection spectroscopy on monoclinic crystals /252/.

Borstel, G., Falge, H.J., Otto, A.: Surface and bulk phonon-polaritons observed by attenuated total reflection /253/.

Agranovich, V.M.: Raman scattering of light by surface polaritons in media with inversion centre /287/.

Wallis, R.F.: Surface-wave phenomena /309/.

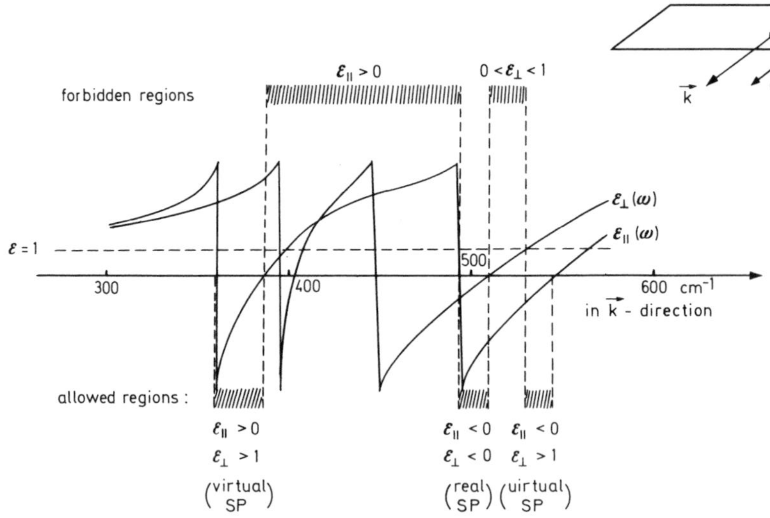

Fig.53 Diagram showing the dielectric functions $\varepsilon_{\perp}(\omega)$ and $\varepsilon_{\parallel}(\omega)$ of α quartz qualitatively. The hatched areas above show regions where no surface polaritons are allowed, whereas in the regions indicated below real and virtual surface polaritons appear. These results correspond to the geometry shown in Fig.52b.

5.5 POLARITON INTERACTION WITH LOCALIZED MODES, SECOND-ORDER PHONONS AND SOFT MODES

Interactions between polaritons and localized modes may advantageously be observed on impurities due to isotopes. When the (natural) abundance of these impurities is of a suitable order, such systems provide good conditions for experimental investigations; maximum homogeneity of the distribution in the sample is guaranteed. Because only an atomic mass is changed, the frequency of the localized mode is moved away from the corresponding fundamental mode by an amount which can frequently be calculated from simple oscillator models. According to 3.1, the local electric field $\underline{E}_{loc}$ (via the electronic polarizabilities) has great influence on the polariton dispersion. Neither the polarizability nor $\underline{E}_{loc}$, however, are essentially changed. The Born-Oppenheimer approximation is justified for such systems. Finally, isotopic impurities cause a type of easily available localized modes. Very recently such experiments on $K_3Cu(CN)_4$ have been reported by Nitsch and Claus /200/. The optical phonon spectrum of this crystal consists of two regions, well separated by more than 1300 cm^{-1}, where no fundamentals exist. The low-frequency region from 0 to about 700 cm^{-1} originates from translational vibrations, librations, and internal vibrations of the K^+ and $Cu(CN)_4^3$ ions. The high-frequency region from 2030 to about 2100 cm^{-1} is caused by $C \equiv N$ stretching vibrations: There are two fundamental A_1(TO) modes at 2074 and 2093 cm^{-1} and one E(TO) phonon at 2080 cm^{-1}, all of which are simultaneously IR- and Raman-active. All the modes show large scattering cross-sections and in fact cause the strongest Raman lines of the entire spectrum. For both the carbon and nitrogen atoms there are stable isotopes, ^{13}C and ^{15}N. The natural abundances are 1.108 % and 0.36 %, respectively. Four modes due to these impurities have been recorded: two of type A_1 at 2032.5 and 2046 cm^{-1}, and two of type E at 2040 and 2053 cm^{-1}. They originate from $^{13}C \equiv N$ and $C \equiv {}^{15}N$ oscillators, respectively. The frequency shifts and relative intensities compared to those of the fundamentals caused by the $^{12}C \equiv {}^{14}N$ oscillators have been calculated and were found to be in good agreement with the experimental data.

Couplings between polaritons associated with the A phonon at 2074 and the E phonon at 2080 cm^{-1} and the isotopic modes have been observed. A representative spectra series is shown in Fig.54a. The scans show four resonance splittings with drastic changes in the intensities. The positions of the maximum resonances and their assignments have been indicated.

Theories for localized polariton modes were published by Mills and Maradudin in 1970 /201/ and by Ohtaka in 1973 /202/. A recent paper by Hwang discussed isotopic effects of polar phonons /203/. Nitsch has carried out numerical calculations based on the dispersion relation derived by Ohtaka and found good agreement with the experimental data from $K_3Cu(CN)_4$ /318/. The dispersion branches shown in Fig.54b facilitate comparison with data corresponding to the spectra in Fig.54a.

When the frequency of a fundamental vibration becomes almost equal

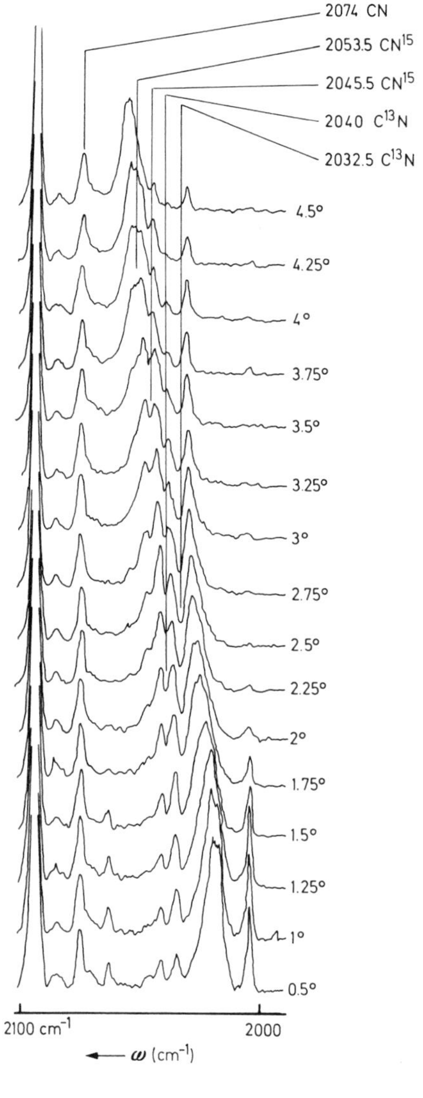

Fig.54a Interactions of polaritons assiciated with the A_1(TO) mode at 2074 cm^{-1} in $K_3Cu(CN)_4$ with four localized modes originating from $^{13}C{\equiv}N$ and $C{\equiv}^{15}N$ vibrations, see text, from /200/.

to that of a second-order phonon, Fermi resonance may take place /13, 14/. Because polaritons can easily be'moved' in the Raman spectra, polariton Fermi resonances can frequently be made to appear. The phenomenon was studied theoretically by Agranovich and

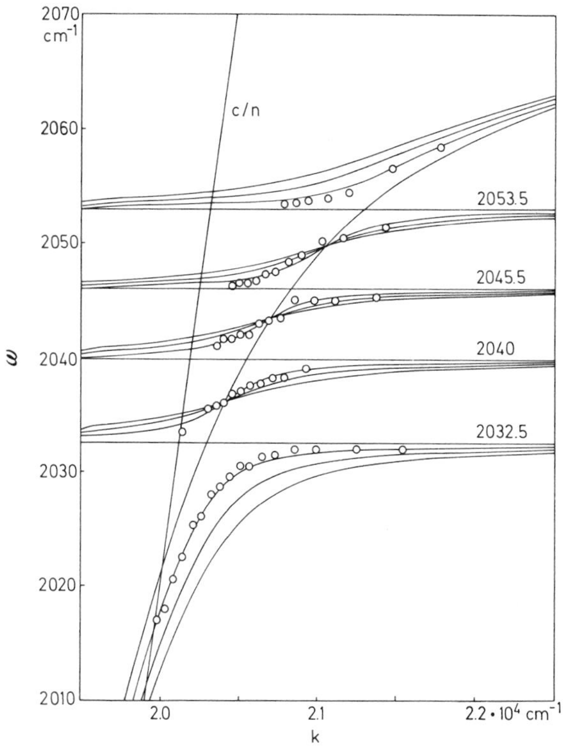

Fig.54b Experimental data and calculated dispersion curves for polariton interaction with localized modes as shown in Fig.54a, from /318/. Three sets of dispersion curves calculated for three different values of the fitting parameter (localized mode amplitudes and oscillator strengths) are shown.

Lalov /204/ in 1971. Experimental results have been reported for orthorhombic α-HIO_3 by Kitaeva et al. /205/. The same group also reported Fermi resonance between a second-order phonon and a polar first-order mode moved by directional dispersion only /132/. Polariton Fermi resonances have also been recorded in $LiNbO_3$ /206, 108/ and $LiIO_3$ /207/. Fig.35 where the polariton branch associated with the E(TO) phonon at 582 cm^{-1} in $LiNbO_3$ 'crosses' a second-order phonon branch at $\sim$ 537 cm^{-1} illustrates the phenomenon.

There are also some experimental studies of polaritons associated with soft modes. The phenomena observed seem to be of special interest because they show behavior typical of polaritons near phase transitions. Data exist for $LiTaO_3$ /208/, $LiNbO_3$ /209/, and $BaTiO_3$ /210-212/. Some representative results are reproduced in Fig.55, showing the temperature dependence of the polariton-dispersion curve associated with the lowest-frequency A_1(TO) mode in $LiNbO_3$, see 4.6. Polaritons in the β phase of quartz have recently been studied by Fries and Claus /323/.

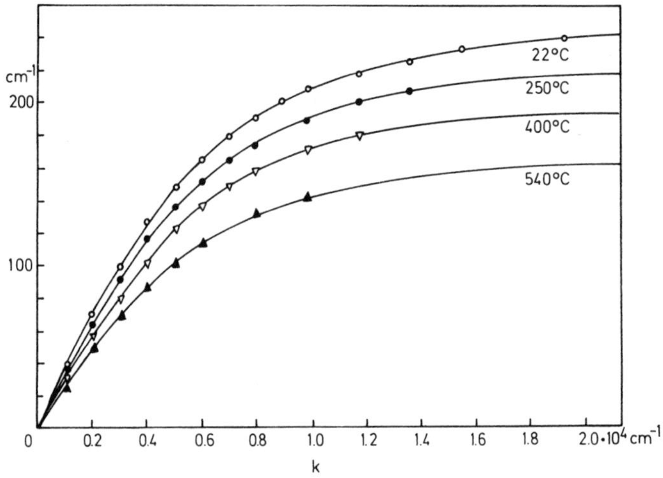

Fig.55 Temperature dependence of the polariton dispersion curve associated with the (soft) A_1(TO) phonon at 255 cm^{-1} in $LiNbO_3$, which was frequently cited in the preceding text, from /209/. (These authors, however, assigned the phonon to appear at 248 cm^{-1} at room temperature.)

Additional Literature

Shepherd, I.W.: Wave-vector-dependent relaxation of an optical phonon in Gadolinium Molybdate /226/.

Appendix 1

THE EWALD METHOD

(Applied to prove the Lorentz relation (4-148) in 4.13)

Apart from providing a proof for (4-148) and (4-149), the Ewald method offers an elegant way to calculate the $B^{C}_{k,k'}(\underline{k})$ and $\gamma_{k,k'}(\underline{k})$ numerically for real crystal lattices /62, 263, 328/.

According to (4-145) and (4-146), the effective field can be written

$$\underline{F}_k(\underline{k}) = -(1/e_k) \sum_{k'} C^{C}_{k,k'}(\underline{k}) \cdot \underline{U}_{k'} \tag{A1-1}$$

where

$$C^{C}_{k,k'}(\underline{k}) = -e_k e_{k'} \exp(i\underline{k}\cdot\underline{R}_{k'}) \sum_{\ell} \frac{\partial^2}{\partial \underline{R}_{k'} \partial \underline{R}_{k'}} \left(\frac{\exp(-i\underline{k}\cdot\underline{R}_{\ell,k})}{|\underline{R}_{\ell,k} - \underline{R}_{k'}|} \right) . \tag{A1-2}$$

The term with $\ell = 0$ has to be excluded for $k = k'$, see (4-144). Because the omission of this term disturbs the calculation, we first generally claim $k \neq k'$. Let the relation

$$1/|\underline{R}_{\ell,k} - \underline{R}_{k'}| = (2/\sqrt{\pi}) \int_0^\infty \exp(-|\underline{R}_{\ell,k} - \underline{R}_{k'}|^2 \rho^2) d\rho \tag{A1-3}$$

be introduced into (A1-2). The sum therein is accordingly replaced by

$$\frac{\partial^2}{\partial \underline{R}_{k'} \partial \underline{R}_{k'}} \int_0^\infty (2/\sqrt{\pi}) \left\{ \sum_{\ell} \exp\left[-|\underline{R}_{\ell,k} - \underline{R}_{k'}|\rho^2 + i\underline{k}\cdot(\underline{R}_{k'} - \underline{R}_{\ell,k}) \right] \right\} \exp(i\underline{k}\cdot\underline{R}_{k'}) d\rho \tag{A1-4}$$

while simultaneously exchanging the summation with the differentiation and integration. The sum over ℓ in (A1-4) can be expanded into a Fourier series

$$\sum_{\ell} \exp[-|\underline{R}_{\ell,k}-\underline{R}_{k'}|\rho^2+i\underline{k}\cdot(\underline{R}_{k'}-\underline{R}_{\ell,k})] =$$

$$= (\sqrt{\pi}^3/\upsilon_a) \sum_h (1/\rho^3)\exp[-|\underline{g}(h)+\underline{k}|/4\rho^2+i\underline{g}(h)\cdot(\underline{R}_k-\underline{R}_{k'})] \quad . \qquad \text{(A1-5)}$$

$$\underline{g}(h) = 2\pi(h_1\underline{b}_1 + h_2\underline{b}_2 + h_3\underline{b}_3) \qquad \text{(A1-6)}$$

are vectors of the reciprocal lattice with $h \equiv (h_1,h_2,h_3)$ covering all combinations of the integers h_1,h_2,h_3. The unit cell is determined by $2\pi\underline{b}_1$, $2\pi\underline{b}_2$, and $2\pi\underline{b}_3$. Eq. (A1-5) has become known as the 'theta transformation'. For convenience, the sum is split into two parts for small and large ρ by introducing a clipping radius r in the reciprocal lattice. By considering the term with h = 0 separately and denoting the remainder by $\sum_h'$, we obtain

$$C^C_{k,k'}(\underline{k}) =$$

$$-e_k e_{k'}\exp(i\underline{k}\cdot\underline{R}_{k'}) \frac{\partial^2}{\partial\underline{R}_{k'}\partial\underline{R}_{k'}} \{(4\pi/\upsilon_a|\underline{k}|^2)\exp[-|\underline{k}|/4r^2-i\underline{k}\cdot\underline{R}_{k'}] +$$

$$+ (\pi/\upsilon_a r^2)\sum_h{}' G[|\underline{g}(h)+\underline{k}|^2/4r^2]\exp[i\underline{g}(h)\cdot(\underline{R}_k-\underline{R}_{k'})-i\underline{k}\cdot\underline{R}_k] +$$

$$+ r\sum_{\ell} H[r|\underline{R}_{\ell\,k,k'}|]\exp[-i\underline{k}\cdot\underline{R}_{\ell,k}]\} \quad . \qquad \text{(A1-7)}$$

The abbreviations used are

$$G(x) = e^{-x}/x \quad ,$$

$$H(x) = (2/x\sqrt{\pi})\int_x^\infty e^{-t^2}\,dt = (1-F(x))/x \quad ,$$

and

$$F(x) = (2/\sqrt{\pi})\int_0^x e^{-t^2}\,dt \qquad \text{(error function).}$$

With

$$\mathcal{H}(\underline{x}) = \frac{\partial^2}{\partial\underline{x}\partial\underline{x}} H(|\underline{x}|) \qquad \text{(A1-8)}$$

where $\underline{x} = r\,\underline{R}_{\ell\,k,k'}$, (A1-7) becomes

$$C^{c}_{k,k'}(\underline{k}) = e_k e_{k'}\Big\{(4\pi/v_a)\,\frac{\underline{k}\,\underline{k}}{|\underline{k}|^2}\exp[-|\underline{k}|^2/4r^2] +$$

$$+(\pi/v_a r^2)\sum_h{}'[\underline{g}(h)+\underline{k}][\underline{g}(h)+\underline{k}]G[|\underline{g}(h)+\underline{k}|/4r^2]\exp[i\underline{g}(h)\cdot(\underline{R}_k-\underline{R}_{k'})] -$$

$$-r^3\sum_\ell H(r\underline{R}_{\ell\,k,k'})\exp[-i\underline{k}\cdot\underline{R}_{\ell\,k,k'}]\Big\} \quad . \qquad \text{(A1-9)}$$

$H(\underline{x})$ is explicitly

$$H(\underline{x}) = [(3/x)(1-F(x))+(6/\sqrt{\pi})\exp(-x^2)/x^4+(4/\sqrt{\pi})\exp(-x^2)/x^2]\,\underline{x}\,\underline{x} -$$

$$-[(1/x^3)(1-F(x))+(2/\sqrt{\pi})\exp(-x^2)/x^2]\,I \quad . \qquad \text{(A1-10)}$$

I denotes the unit matrix and $x = |\underline{x}| = r|\underline{R}_{\ell\,k,k'}|$. As has already been pointed out above, (A1-9) holds only for $k \neq k'$. The difficulty in the summation for $k = k'$ originates from the fact that no term with $\ell = 0$ appears in (A1-2) or (4-145). The Coulomb interaction of an ion with itself has to be excluded. Consequently there is no term with $\ell=\ell'$ and $k=k'$ in (4-144). The Ewald method, however, may also be applied in this case by claiming $\underline{R}_k \neq \underline{R}_{k'}$ and formally including the ($\ell = 0$)-term in the summation. This term has later to be subtracted from the result and the relation obtained will be considered in the limit $\underline{R}_k \to \underline{R}_{k'}$. Formally H(x) in the $\ell = 0$ term is replaced by

$$H^{o}(x) = -(2/x\sqrt{\pi})\int_0^x \exp(-x^2)\,dx \quad . \qquad \text{(A1-11)}$$

For details of the corresponding derivation we refer to /62/.

Eq.(A1-9) can advantageously be used for numerical calculations of the $C^{c}_{k,k'}(\underline{k})$ because all partial sums rapidly converge for a suitable clipping radius r. On the other hand, terms corresponding to a 'macroscopic electric field' can now easily be separated from (A1-9). The sums $\sum_h{}'$ and $\sum_\ell$ are also regular for $\underline{k} = 0$. The term

$$(4\pi/v_a)\,\frac{\underline{k}\,\underline{k}}{|\underline{k}|^2}\exp[-|\underline{k}|^2/4r^2] = (4\pi/v_a)\,\frac{\underline{k}\,\underline{k}}{|\underline{k}|^2}\Big(1-\frac{|\underline{k}|^2}{4r^2}+\ldots\Big)\ , \qquad \text{(A1-12)}$$

however, is singular because $\underline{k}\,\underline{k}/|\underline{k}|^2$ behaves discontinuously for $\underline{k} = 0$. By isolating this term and introducing the following abbreviation for the rest

$$B_{k,k'}(\underline{k}) = e_k e_{k'}\Big\{(4\pi/\upsilon_a)\,\frac{\underline{k}\,\underline{k}}{|\underline{k}|^2}\left[\exp(-|\underline{k}|^2/4r^2)-1\right] -$$

$$-r^3 \sum_{\ell} H(r\,\underline{R}_{\ell}^{\,k,k'})\exp(-i\underline{k}\cdot\underline{R}_{\ell}^{\,k,k'}) +$$

$$+(\pi/\upsilon_a r^2)\sum_{h}{}'\left[\underline{g}(h)+\underline{k}\right]\left[\underline{g}(h)+\underline{k}\right] G\left[|\underline{g}(h)+\underline{k}|^2/4r^2\right]\exp\left[i\underline{g}(h)\cdot(\underline{R}_k-\underline{R}_{k'})\right]\Big\} \quad , \tag{A1-13}$$

we can write (A1-9)

$$C^{C}_{k,k'}(\underline{k}) = (4\pi/\upsilon_a)e_k e_{k'}\,\frac{\underline{k}\,\underline{k}}{|\underline{k}|^2} + B_{k,k'}(\underline{k}) \quad . \tag{A1-14}$$

The effective field (4-146) therefore becomes

$$\underline{F}_k(\underline{k}) = -\sum_{k'}(4\pi/\upsilon_a)\,\frac{\underline{k}\,\underline{k}}{|\underline{k}|^2}\cdot\underline{U}_{k'} - \sum_{k'}(1/e_k)B_{k,k'}(\underline{k})\cdot\underline{U}_{k'} \quad . \tag{A1-15}$$

The first term in (A1-15), being independent of sublattice k, describes the macroscopic field

$$\underline{E}(\underline{k}) \equiv -4\pi\,\frac{\underline{k}\,\underline{k}}{|\underline{k}|^2}\cdot\sum_{k'}(e_{k'}/\upsilon_a)\underline{U}_{k'} \quad . \tag{A1-16}$$

Using this abbreviation, we can rewrite (A1-15)

$$\underline{F}_k(\underline{k}) = \underline{E}(\underline{k}) - \sum_{k'}(1/e_k)B_{k,k'}(\underline{k})\cdot\underline{U}_{k'} \quad , \tag{A1-17}$$

which is obviously identical to (4-148). It is readily seen that (4-149) thus holds too.

$\underline{E}(\underline{k})$ represents the electric field derived from Maxwell's equations in the electrostatic limit ($n^2 \to \infty$). By introducing $\underline{P} = \sum_{k'}(e_{k'}/\upsilon_a)\underline{U}_{k'}$ into (A1-16) and taking into account $\underline{s} = \underline{k}/|\underline{k}|$, we get

$$\underline{E}(\underline{k}) = -4\pi\,\underline{s}(\underline{s}\cdot\underline{P}) \quad . \tag{A1-18}$$

This result directly corresponds to (3-19) or (4-9) for $n^2 \to \infty$.

Appendix 2

THE MICROSCOPIC TREATMENT BY PICK

In 3.4 we showed that Huang's equations can be derived from a potential energy density

$$\Phi = -(1/2)\underline{Q}\cdot B^{11}\cdot\underline{Q} - \underline{Q}\cdot B^{12}\cdot\underline{E} - (1/2)\underline{E}\cdot B^{22}\cdot\underline{E} \quad . \qquad \text{(A2-1)}$$

Macroscopic considerations presented by Pick /80/ start from a similar relation which, however, includes the vacuum field $(1/8\pi)\underline{E}^2$. The nomenclature used by Pick is somewhat different from ours. In order better to show the analogies, we therefore translate the most important formulas into our language. Numbers in curly brackets refer to the equations in Pick's article.

The energy density {1} is introduced as

$$\Phi = -(1/2)\underline{Q}\cdot B^{11}\cdot\underline{Q} - \underline{E}\cdot B^{21}\cdot\underline{Q} - (1/2)\underline{E}\cdot B^{22}\cdot\underline{E} - (1/8\pi)\underline{E}^2 \quad . \qquad \text{(A2-2)}$$

Eqs.{5a} and {6a} stand for

$$-\omega^2\underline{Q} = -(\partial\Phi/\partial\underline{Q}) \quad . \qquad \text{(A2-3)}$$

The equation of motion is derived from (A2-2) and (A2-3). The use of the more general expression for the energy density, however, means that the electric displacement $\underline{D}$ is obtained instead of the polarization $\underline{P}$ when $-4\pi\partial\Phi/\partial\underline{E}$ is applied to (A2-2). This can easily be verified:

$$4\pi(B^{21}\cdot\underline{Q} + B^{22}\cdot\underline{E}) + \underline{E} = 4\pi\underline{P} + \underline{E} = \underline{D} \quad . \qquad \text{(A2-4)}$$

Maxwell's equations are summarized in {6b} to

$$\underline{D} = n^2(I-\underline{s}\ \underline{s})\cdot\underline{E} \tag{A2-5}$$

where $\underline{s}\ \underline{s}$ stands for a dyad, see (3-17). When (A2-5) is combined with the relation for the displacement derived from the energy density, Huang's equations appear together with Maxwell's equations as

$$(-B^{11}-I\omega^2)\cdot\underline{Q}-(B^{21})^{+}\cdot\underline{E} = 0 \tag{A2-6}$$

$$-B^{21}\cdot\underline{Q}+(1/4\pi)\left[n^2(I-\underline{s}\ \underline{s})-(4\pi B^{22}+I)\right]\cdot\underline{E} = 0 \quad . \tag{A2-7}$$

I denotes the unit matrix. A comparison of these equations with the corresponding set in Pick's article {9} allows ready identification of the different symbols used there. The electric field $\underline{E}$ can be eliminated from (A2-6) and (A2-7). The remaining equation of motion contains only the mechanical displacements $\underline{Q}$. The condition for non-trivial solutions is that the determinant of the dynamical matrix vanishes

$$\left|-B^{11}-4\pi(B^{21})^{+}\left[n^2(I-\underline{s}\ \underline{s})-(4\pi B^{22}+I)\right]^{-1}B^{21}-\omega^2 I\right| = 0 \quad . \tag{A2-8}$$

As a result, the generalized Fresnel equation is obtained in the same way as in 4.2. (A2-8) is identical with {43} in Pick's article. Pick has left the coefficients $B^{\mu\nu}$ $\underline{k}$-dependent; this difference, however, is of importance only for $\underline{k} \gtrsim 10^7\ \text{cm}^{-1}$, see 4.13.

An equation equivalent to (A2-8) has then been derived from microscopic properties of the crystal. We summarize some aspects of the derivation using in part a simplified notation to allow easier comparison with the microscopic treatment in 4.13. For details of the derivation we refer to the original paper.

The discussion is based on a hamiltonian for the neutral system of electrons and nuclei. For a fixed position of the latter (Born-Oppenheimer approximation)

$$H_o = \sum_a \frac{P_a}{2M_a} + (1/2)\sum_{\substack{a,b\\ a\neq b}} \frac{(Z_a e)(Z_b e)}{|\underline{R}_a-\underline{R}_b|} +$$

$$+ \sum_t \frac{P_t}{2m} - \sum_{a,t}\frac{(Z_a e)e}{|\underline{R}_a-\underline{r}_t|} + (1/2)\sum_{\substack{t,u\\ t\neq u}}\frac{e^2}{|\underline{r}_t-\underline{r}_u|} \quad . \tag{A2-9}$$

The five groups of terms denote respectively (1) the kinetic energy of the nuclei (M_a = masses, P_a = momenta), (2) the potential energies of the nuclei a and b with respect to each other (Ze = effective charges, e = elementary-charge, $\underline{R}$ = position vectors), (3) the kinetic energy of the electrons (m = mass, P_t = momenta), (4) the potential energy of the electrons with respect to the nuclei ($\underline{r}$ = electron position vectors), and (5) the potential energy of the electrons t and u with respect to each other.

The nuclei have been treated classically. They are assumed to have an equilibrium position $\underline{R}_a^o$. The electric field at another position $\underline{R}_a$ can be derived from the second and fourth terms in (A2-9) by making use of the Helman-Feynman theorem

$$\underline{E}(R_a) = -\frac{\partial}{\partial \underline{R}_a}\left(\sum_b \frac{Z_b e}{|\underline{R}_a - \underline{R}_b|} + \int \frac{\langle\rho(\underline{r})\rangle}{|\underline{R}_a - \underline{r}|} d^3\underline{r}\right) \quad . \qquad \text{(A2-10)}$$

The sum over all individual electrons has been replaced by an integral over space. $\langle\rho(\underline{r})\rangle$ stands for the mean electronic charge density which must be introduced because the electrons cannot be treated classically. The equation of motion for a nucleon with the position vector $\underline{R}_a$ is derived using the classical Lorentz force

$$M_a \ddot{\underline{u}}_a = Z_a e E(\underline{R}_a) = Z_a e\left[\underline{E}(\underline{R}_a^o) + \sum_b \int_{-\infty}^{t} (\partial \underline{E}/\partial \underline{R}_b^o)\underline{u}_b \, dt'\right] \quad . \qquad \text{(A2-11)}$$

$\underline{E}(\underline{R}_a)$ has been expanded into a series in terms of the time-dependent displacement vectors $\underline{u}_b$ in the equilibrium position. The expansion is restricted to the harmonic approximation. According to (A2-10) the electric field $\underline{E}(\underline{R}_a^o)$ vanishes. $\underline{E}$ has to be substituted only in the second term on the right-hand side of (A2-11), {53}, using the relation (A2-10), {49b}. The dynamical matrix of the equation of motion obtained in this way contains terms that depend on the positions of the nuclei as well as a term depending on the electronic charge density. The eigenvalues and eigenvectors describe the pure phonon modes.

A description of polariton modes is achieved by introducing the classic wave equation {62} in the form

$$\text{curl curl } \underline{E} + (1/c^2)\ddot{\underline{E}} = -(4\pi/c^2)(\partial \underline{j}/\partial t) \quad . \qquad \text{(A2-12)}$$

$\underline{j}$ denotes the current density which in general consists of an externally applied part $\underline{j}_{ext}$ and an induced electronic part $\underline{j}_{el}$. The latter can be interpreted as the polarization current

$$\underline{j}_{el} = \partial\underline{P}/\partial t \quad . \qquad \text{(A2-13)}$$

Insulators may be described by taking only (A2-13) into account. Pick's derivation is broad enough to include the existence of plasmons.

$\underline{j}_{el}$ is shown to be correlated to the total electric field by a linear response function. On the other hand, the continuity equation correlates the current density to the charge density. Thus the wave equation (A2-12), {62}, is coupled with the equation of motion via the current much as Maxwell's equations were earlier coupled with the equation of motion in the macroscopic treatment via the polarization, see (A2-13) and Chapters 3 and 4. The dynamical matrix of the coupled system has been derived as {93} or {122} in which the form becomes identical to (A2-8), {43}. Finally relations for the coefficients B^{11}, B^{21} and $(4\pi B^{22}+I)$ are derived in {86}, {90}, and {88}, respectively. In principle, they allow the coefficients to be calculated from microscopic data. The difference from the result in the macroscopic treatment is that the coefficients remain dependent on both wave-vector and frequency, just as they turned out to be in 4.13. Pick, however, was also able to show that they can be regarded as constants around the center of the 1BZ, which is the real polariton region.

Appendix 3

THE RESPONSE FUNCTION TREATMENT BY BARKER AND LOUDON

This appendix is not intended to be a complete review of the very extensive article by Barker and Loudon /73/. We merely point out some aspects in order to facilitate comparison of their derivations with our text.

The power of radiation from an oscillating dipole was related to the mean square of the dipole moment by (1-3). The oscillating dipole moments themselves are caused by thermal fluctuations of the microscopic ionic and electronic oscillators in the material. Inelastic scattering of light from polaritons and polar phonons in ideal crystals is thus primarily due to these fluctuations. A detailed discussion of fluctuations on the basis of thermodynamics and statistics has been given by Landau and Lifschitz /Ref.304, Sections 112-127/.

Let us consider the relation

$$(u^2)_\omega = (kT\,\Gamma/\pi)\left[m^2(\omega_0^2-\omega^2)^2+\omega^2\Gamma^2\right]^{-1} \quad , \tag{A3-1}$$

/Ref.304, Eq.(124,16)/ where k is the Boltzmann constant, Γ the damping factor, m the oscillator mass, and ω_0 the eigenfrequency of the system. The mean square of the displacement coordinate u at the frequency ω has been related to the temperature of the system. More generally, this relation can be written

$$(u^2)_\omega = (kT/\pi\omega)\,\mathrm{Im}\;\phi \tag{A3-2}$$

where ϕ has become known as the response function of the system. ϕ is explicitly

$$\phi = \left[m(\omega_o^2-\omega^2)-i\omega\Gamma\right]^{-1} \quad . \tag{A3-3}$$

An equation equivalent to (A3-2) was derived by Nyquist in 1928 /307/. His equation determines the mean square of the voltage caused by thermal fluctuations in a conductor, i.e. the temperature. The response function of the system is there the impedance.

Eq.(A3-2), generally known as Nyquist's theorem, is valid only in the high-temperature limit. For low temperature rigorously

$$(u^2)_\omega = (\hbar/\pi)\left[1+(\exp(\hbar\omega/kT)-1)^{-1}\right]\mathrm{Im}\,\phi \quad . \tag{A3-4}$$

A derivation of (A3-4) can again be found in /Ref.304, Eq.(126,9)/, but there the thermal factor is somewhat different because a symmetric combination of the Fourier components at $+\omega$ and $-\omega$ has been considered, see /73, comment 2/. The mean square of the time-dependent fluctuating displacement coordinate u(t) over the whole frequency spectrum finally becomes

$$\langle u^2(t)\rangle = \int_{-\infty}^{+\infty}(u^2)_\omega \, d\omega \quad , \tag{A3-5}$$

/304, Eq.(121,6)/. Barker and Loudon could show that (A3-4) correctly describes the 'intensities' of the Stokes and Anti-Stokes components at $+\omega_o$ and $-\omega_o$, respectively, see Fig.1 of their paper. Note, however, when comparing their Figs.1c and 2b, that the Stokes peak corresponds to a phonon creation process described by $+\omega_o$ whereas the Raman spectrum shows a corresponding shift of $-\omega_o$ with respect to the exciting laser frequency.

Regarding the form of the response function (A3-3), it can easily be verified that for the mechanical system in question ϕ can be expressed by

$$\phi(\omega) = u(t)/F(t) \tag{A3-6}$$

where F(t) denotes an external force driving the damped oscillator. The ions and electrons in crystals are treated as such oscillators driven 'externally' by an electric field $\underline{E}$. The corresponding equations of motion for the simplest case of a cubic diatomic lattice have been introduced as

$$M \ddot{U} + K U = Z E \tag{A3-7}$$

$$m \ddot{u} + k u = z E \quad . \tag{A3-8}$$

Capital letters refer to ions and small letters to electrons. Z and z denote charges and K and k are force constants. The electric field $\underline{E}$ introduced in these equations has strictly to be interpreted as the local electric field from 3.1. Throughout the derivations by Barker and Loudon, however, it is frequently interpreted also as the macroscopic field. Their text has to be read with some care in view of this aspect. In addition to (A3-7) and (A3-8), the polarization per unit volume V is introduced

$$P = (Z U + z u)/V \quad , \tag{A3-9}$$

see {2.1} to {2.3}. (Curly brackets refer to the equations of Barker and Loudon.)

This is the way Huang's equations have been introduced by the authors. Furthermore, Maxwell's equations are introduced in the usual way in order to describe the coupling with photons. From (A3-7) to (A3-9) the mode strengths {2.10}, the high-frequency dielectric constant {2.13}, and the frequency-dependent dielectric function {2.9}, {2.12} are derived. The dielectric function is correlated to the linear response function which consequently may be regarded as the origin of polariton dispersion. The model has been generalized to multiatomic, nonisotropic materials. The results obtained agree with those derived in our preceding text.

An extension to include nonlinear effects has been achieved by replacing (A3-7) and (A3-8) by

$$\left.\begin{aligned} M \ddot{U} + K U + \alpha(U-u)^2 &= Z E \\ m \ddot{u} + k u + \alpha(U-u)^2 &= z E \end{aligned}\right. \quad . \tag{A3-10}$$

This implies that nonlinear coupling between the ions and electrons is taken into account, {2.28} and {2.29}.

Scattering cross-sections of polaritons have been derived in {2.58} and {3.71} taking into account Nyquist's theorem (A3-2).

Finally, resonance effects are discussed in the way that the exciting laser frequency is supposed to approach the electronic frequency $(k/m)^{1/2}$, see (A3-8).

Appendix 4

RAMAN TENSOR TABLES FOR THE 32 CRYSTAL CLASSES

Tables for the symmetric parts of the Raman tensors have been published by several authors /31, 33, 56, 58, and others/. Because most of these tables contain some errors, we include a corrected one. The following properties are listed in the different columns:

1. orientation of the principal axes in agreement with those used by Nye /Ref.45, Appdx B, Table 21/
2. Schoenfliess and international symbols for the crystal classes
3. optical activity /342/
4. decomposition of a polar second-rank tensor which is not claimed to be symmetric, see 2.5 and 2.6. The Wigner-Eckart theorem /343/ has not been applied. (O) denotes the zero matrix.

The index combinations $\rho\sigma$ simply indicate the Raman tensor elements $S_{\rho\sigma}$ $(\rho,\sigma=x,y,z)$.

In the trigonal, tetragonal, and hexagonal crystal classes

$a = (1/2)(xx+yy) \equiv (1/2(S_{xx}+S_{yy})$,
$b = (1/2)(xy-yx)$,
$c = (1/2)(xx-yy)$,
$d = (1/2)(xy+yx)$,

and in the cubic crystal classes

$a = (1/3)(xx+yy+zz)$,
$b = (1/6)(yy-zz)$,
$c = (1/6)(zz-xx)$,
$d = (1/6)(xx-yy)$,
$e = (1/2)(yz-zy)$,
$f = (1/2)(yz+zy)$,
$g = (1/12\sqrt{2})(xx+yy-2zz)$, $h = (\sqrt{3}/12\sqrt{2})(xx-yy)$.

The cartesian coordinate ρ given in brackets after some of the symmetry-species symbols indicates infrared activity due to the component P_ρ of the electric dipole moment. The translation T_ρ is simultaneously allowed. A component R_ρ correspondingly shows that the rotation is allowed. According to (2-31), the selection rules for Raman tensors are given in this way.

Due to (1-7), (1-8), and (4-124), the relative scattering intensity of the m-th <u>nonpolar phonon</u> of a certain symmetry species γ can be calculated from

$$I \sim \sum_{j=1}^{d_\gamma} \left| (e_i^x, e_i^y, e_i^z) \begin{pmatrix} {}^{(j)} \\ S^{(m)}_{\rho\sigma} \end{pmatrix} \begin{pmatrix} e_s^x \\ e_s^y \\ e_s^z \end{pmatrix} \right|^2 . \tag{A4-1}$$

The summation index j covers the degenerated irreducible representations, d_γ being the degree of degeneration. $\underline{e}_i$ and $\underline{e}_s$ denote unit vectors in direction of the polarization of the incident and scattered light, respectively. The absolute value signs are of importance because the Raman tensors may be used also in their complex forms.

For long-wavelength <u>polar phonons</u> in the cubic crystal classes we have

$$I \sim \left| (e_i^x, e_i^y, e_i^z) \left\{ (S^{(m)}_{x,\rho\sigma}, S^{(m)}_{y,\rho\sigma}, S^{(m)}_{z,\rho\sigma}) \begin{pmatrix} Q_x^{(m)} \\ Q_y^{(m)} \\ Q_z^{(m)} \end{pmatrix} \right\} \begin{pmatrix} e_s^x \\ e_s^y \\ e_s^z \end{pmatrix} \right|^2 . \tag{A4-2}$$

The $S^{(m)}_{x,\rho\sigma}$, $S^{(m)}_{y,\rho\sigma}$ and $S^{(m)}_{z,\rho\sigma}$ are those introduced in (2-54). When examining a certain symmetry species only the polar components of $S^{(m)}$ appearing in this species have to be considered. The normal coordinate vector $\underline{Q}^{(m)}$ describes the mechanical polarization of the m-th mode. The tensor in curly brackets, frequently referred to as the 'effective scattering tensor', is identical to the first sum in (4-126) for purely transverse modes. The $S^{(m)}_{\tau,\rho\sigma}$ are identical with the atomic displacement tensor components $a^{(m)}_{\alpha,\beta\gamma}$, see (4-129).

For long-wavelength polar (extraordinary) phonons in arbitrary directions, polar LO phonons, and polaritons generally, the electric field $\underline{E}$ does not vanish. As a result the electro-optic terms in (4-126) may make an important contribution, and all normal coordinates are consequently coupled. The relative scattering intensities in these cases can be calculated only from (4-124) with the use of (4-126). The electro-optic tensor ($b_{\alpha,\beta\gamma}$) has the same symmetry properties as the atomic displacement tensor. Its form can therefore be seen from the table. Loudon reports a formula for the scattering intensity where both tensors have been combined /Ref.31, Eq.(53)/. This relation, although frequently cited, holds only for diatomic uniaxial crystals and for the principal directions in polyatomic crystals.

1	2	3	4
	Crystal Class	Optical Activity	Raman tensors S and S', resp.; see 2.6 and /31, 33, 56, 57/
Triclinic			$\begin{pmatrix} xx & xy & xz \\ yx & yy & yz \\ zx & zy & zz \end{pmatrix}$ (O)
	$C_1 = 1$	yes	$A(x,y,z;R_x,R_y,R_z)$
	$C_i = \bar{1}$	no	$A_g(R_x,R_y,R_z)$ $A_u(x,y,z)$
Monoclinic			$\begin{pmatrix} xx & & xz \\ & yy & \\ zx & & zz \end{pmatrix}$ $\begin{pmatrix} & xy & \\ yx & & yz \\ & zy & \end{pmatrix}$ (O) (O)
$C_2 \parallel y$	$C_2 = 2$	yes	$A(y,R_y)$ $B(x,z;R_x,R_z)$
$\sigma_h \perp y$	$C_s = m$	yes	$A'(x,z;R_y)$ $A''(y;R_x,R_z)$
$C_2 \parallel y$	$C_{2h} = 2/m$	no	$A_g(R_y)$ $B_g(R_x,R_z)$ $A_u(y)$ $B_u(x,z)$
Orthorhombic			$\begin{pmatrix} xx & & \\ & yy & \\ & & zz \end{pmatrix}$ $\begin{pmatrix} & xy & \\ yx & & \\ & & \end{pmatrix}$ $\begin{pmatrix} & & xz \\ & & \\ zx & & \end{pmatrix}$ $\begin{pmatrix} & & \\ & & yz \\ & zy & \end{pmatrix}$ (O) (O) (O)
$C_2^\rho \parallel \rho$	$D_2 = 222$	yes	A $B_1(z;R_z)$ $B_2(y;R_y)$ $B_3(x;R_x)$
$C_2^z \parallel z; \sigma_y \parallel x$	$C_{2v} = mm2$	yes	$A_1(z)$ $A_2(R_z)$ $B_1(x;R_y)$ $B_2(y;R_x)$
$C_2^z \parallel z; \sigma_y \parallel x$	$D_{2h} = mmm$	no	A_g $B_{1g}(R_z)$ $B_{2g}(R_y)$ $B_{3g}(R_x)$ $B_{1u}(z)$ $B_{2u}(y)$ $B_{3u}(x)$

1	2	3	4
Trigonal			$\begin{pmatrix} a & b & \\ -b & a & \\ & & zz \end{pmatrix}$ $\sqrt{2}\begin{pmatrix} c & d & xz/2 \\ d & -c & yz/2 \\ zx/2 & zy/2 & \end{pmatrix}$ $\sqrt{2}\begin{pmatrix} d & -c & -yz/2 \\ -c & -d & xz/2 \\ -zy/2 & zx/2 & \end{pmatrix}$ (O) (O)
$C_3 \parallel z$	$C_3 = 3$	yes	$A(z;R_z)$ $E(x;R_x,R_y)$ $E(y;R_x,R_y)$
$C_3 \parallel z$	$C_{3i} = \bar{3}$	no	$A_g(R_z)$ $E_g^{(1)}(R_x,R_y)$ $E_g^{(2)}(R_x,R_y)$ $A_u(z)$ $E_u(x,y)$
			$\begin{pmatrix} a & & \\ & a & \\ & & zz \end{pmatrix}$ $\begin{pmatrix} & b & \\ -b & & \\ & & \end{pmatrix}$ $\begin{pmatrix} c & & \\ & -c & yz \\ & zy & \end{pmatrix}$ $\begin{pmatrix} & -c & -yz \\ -c & & \\ -zy & & \end{pmatrix}$ (O) (O)
$C_3 \parallel z; C_2 \parallel x$	$\mathcal{D}_3 = 32$	yes	A_1 $A_2(z;R_z)$ $E(x;R_x)$ $E(y{=}x;R_y{=}R_x)$
$C_3 \parallel z; C_2 \parallel x$	$\mathcal{D}_{3d} = \bar{3}m$	no	A_{1g} $A_{2g}(R_z)$ $E_g^{(1)}(R_x)$ $E_g^{(2)}(R_y{=}R_x)$ $A_{2u}(z)$ $E_u(x,y)$
			$\begin{pmatrix} a & & \\ & a & \\ & & zz \end{pmatrix}$ $\begin{pmatrix} & b & \\ -b & & \\ & & \end{pmatrix}$ $\begin{pmatrix} & d & xz \\ d & & \\ zx & & \end{pmatrix}$ $\begin{pmatrix} d & & \\ & -d & xz \\ & zx & \end{pmatrix}$
$C_3 \parallel z; \sigma_v \parallel y$	$C_{3v} = 3m$	no	$A_1(z)$ $A_2(R_z)$ $E(x;R_y)$ $E(y{=}x;R_x{=}R_y)$

1	2	3	4							
Tetragonal			$\begin{pmatrix} a & b & \\ -b & a & \\ & & zz \end{pmatrix}$	$\begin{pmatrix} c & d & \\ d & -c & \\ & & \end{pmatrix}$	$\frac{1}{\sqrt{2}}\begin{pmatrix} & & xz \\ & & yz \\ zx & zy & \end{pmatrix}$	$\frac{1}{\sqrt{2}}\begin{pmatrix} & & -yz \\ & & xz \\ -zy & zx & \end{pmatrix}$	(O)	(O)		
$C_4 \parallel z$	$\mathcal{C}_4 = 4$	yes	$A(z;R_z)$	B	$E(x;R_x,R_y)$	$E(y;R_y=R_x)$				
$S_4 \parallel z$	$\mathcal{S}_4 = \bar{4}$	yes	$A(R_z)$	$B(z)$	$E(x;R_x,R_y)$	$E(-y;R_y=R_x)$				
$C_4 \parallel z$	$\mathcal{C}_{4h} = 4/m$	no	$A_g(R_z)$	B_g	$E_g^{(1)}(R_x,R_y)$	$E_g^{(2)}(R_y=R_x)$	$A_u(z)$	$E_u(x,y)$		
			$\begin{pmatrix} a & & \\ & a & \\ & & zz \end{pmatrix}$	$\begin{pmatrix} & b & \\ -b & & \\ & & \end{pmatrix}$	$\begin{pmatrix} c & & \\ & -c & \\ & & \end{pmatrix}$	$\begin{pmatrix} & d & \\ d & & \\ & & \end{pmatrix}$	$\begin{pmatrix} & & xz \\ & & \\ zx & & \end{pmatrix}$	$\begin{pmatrix} & & \\ & & xz \\ & zx & \end{pmatrix}$		
$C_4 \parallel z; \sigma_v \parallel x$	$\mathcal{C}_{4v} = 4mm$	no	$A_1(z)$	$A_2(R_z)$	B_1	B_2	$E(x;R_y)$	$E(y=x;R_x=R_y)$		
			$\begin{pmatrix} a & & \\ & a & \\ & & zz \end{pmatrix}$	$\begin{pmatrix} & b & \\ -b & & \\ & & \end{pmatrix}$	$\begin{pmatrix} c & & \\ & -c & \\ & & \end{pmatrix}$	$\begin{pmatrix} & d & \\ d & & \\ & & \end{pmatrix}$	$\begin{pmatrix} & & \\ & & yz \\ & zy & \end{pmatrix}$	$\begin{pmatrix} & & -yz \\ & & \\ -zy & & \end{pmatrix}$	(O)	(O)
$C_4 \parallel z; C_2' \parallel x$	$\mathcal{D}_4 = 422$	yes	A_1	$A_2(z;R_z)$	B_1	B_2	$E(x;R_x)$	$E(y=x;R_y=R_x)$		
$S_4 \parallel z; C_2' \parallel x$	$\mathcal{D}_{2d} = \bar{4}2m$	yes	A_1	$A_2(R_z)$	B_1	$B_2(z)$	$E(x;R_x)$	$E(-y=-x;R_y=R_x)$		
$C_4 \parallel z; C_2' \parallel x$	$\mathcal{D}_{4h} = 4mmm$	no	A_{1g}	$A_{2g}(R_z)$	B_{1g}	B_{2g}	$E_g^{(1)}(R_x)$	$E_g^{(2)}(R_y=R_x)$	$A_{2u}(z)$	$E_u(x,y)$

1	2	3	4							
Hexagonal			$\begin{pmatrix} a & b & \\ -b & a & \\ & & zz \end{pmatrix}$	$\frac{1}{\sqrt{2}}\begin{pmatrix} & & xz \\ & & yz \\ zx & zy & \end{pmatrix}$	$\frac{1}{\sqrt{2}}\begin{pmatrix} & & -yz \\ & & xz \\ -zy & zx & \end{pmatrix}$	$\frac{1}{\sqrt{2}}\begin{pmatrix} c & d & \\ d & -c & \\ & & \end{pmatrix}$	$\frac{1}{\sqrt{2}}\begin{pmatrix} -c & -d & \\ -c & -d & \\ & & \end{pmatrix}$	(O)	(O)
$C_6 \parallel z$	$C_6 = 6$	yes	$A(z;R_z)$	$E_1(x;R_x,R_y)$	$E_1(y;R_x=R_y)$	$E_2^{(1)}$	$E_2^{(2)}$		
$C_3 \parallel z$	$C_{3h} = \bar{6}$	no	$A'(R_z)$	$E''^{(1)}(R_x,R_y)$	$E''^{(2)}(R_x=R_y)$	$E'(x)$	$E'(y)$	$A''(z)$	
$C_6 \parallel z$	$C_{6h} = 6/m$	no	$A_g(R_z)$	$E_{1g}^{(1)}(R_x,R_y)$	$E_{1g}^{(2)}(R_x=R_y)$	$E_{2g}^{(1)}$	$E_{2g}^{(2)}$	$A_u(z)$	$E_{1u}(x,y)$
			$\begin{pmatrix} a & & \\ & a & \\ & & zz \end{pmatrix}$ $\begin{pmatrix} & b & \\ -b & & \\ & & \end{pmatrix}$	$\begin{pmatrix} & & \\ & & yz \\ & zy & \end{pmatrix}$	$\begin{pmatrix} & & -yz \\ & & \\ -zy & & \end{pmatrix}$	$\begin{pmatrix} & -c & \\ -c & & \\ & & \end{pmatrix}$	$\begin{pmatrix} c & & \\ & -c & \\ & & \end{pmatrix}$	(O)	(O)
$C_6 \parallel z; C_2' \parallel x$	$D_6 = 622$	yes	A_1 $A_2(z;R_z)$	$E_1(x;R_x)$	$E_1(y=x;R_y=R_x)$	$E_2^{(1)}$	$E_2^{(2)}$		
$C_2 \parallel z; C_2' \parallel x$	$D_{6h} = 6/mmm$	no	A_{1g} $A_{2g}(R_z)$	$E_{1g}^{(1)}(R_x)$	$E_{1g}^{(2)}(R_y=R_x)$	$E_{2g}^{(1)}$	$E_{2g}^{(2)}$	$A_{2u}(z)$	$E_{1u}(x,y)$
			$\begin{pmatrix} a & & \\ & a & \\ & & zz \end{pmatrix}$ $\begin{pmatrix} & b & \\ -b & & \\ & & \end{pmatrix}$	$\begin{pmatrix} & & \\ & & yz \\ & zy & \end{pmatrix}$	$\begin{pmatrix} & & -yz \\ & & \\ -zy & & \end{pmatrix}$	$\begin{pmatrix} & d & \\ d & & \\ & & \end{pmatrix}$	$\begin{pmatrix} d & & \\ & -d & \\ & & \end{pmatrix}$	(O)	
$C_3 \parallel z; C_2 \parallel y$	$D_{3h} = \bar{6}m2$	no	A_1' $A_2'(R_z)$	$E''^{(1)}(R_x)$	$E''^{(2)}(R_y=R_x)$	$E'(x)$	$E'(y=x)$	$A_2''(z)$	
			$\begin{pmatrix} a & & \\ & a & \\ & & zz \end{pmatrix}$ $\begin{pmatrix} & b & \\ -b & & \\ & & \end{pmatrix}$	$\begin{pmatrix} & & xz \\ & & \\ zx & & \end{pmatrix}$	$\begin{pmatrix} & & \\ & & xz \\ & zx & \end{pmatrix}$	$\begin{pmatrix} & -c & \\ -c & & \\ & & \end{pmatrix}$	$\begin{pmatrix} c & & \\ & -c & \\ & & \end{pmatrix}$		
$C_6 \parallel z; \sigma_v \parallel x$	$C_{6v} = 6mm$	no	$A_1(z)$ $A_2(R_z)$	$E_1(x;R_y)$	$E_1(y=x;R_x=R_y)$	$E_2^{(1)}$	$E_2^{(2)}$		

1	2	3	4
Cubic			$\begin{pmatrix} a & & \\ & a & \\ & & a \end{pmatrix} \begin{pmatrix} g+\sqrt{3}h & & \\ & g-\sqrt{3}h & \\ & & -2g \end{pmatrix} \begin{pmatrix} h-\sqrt{3}g & & \\ & h+\sqrt{3}g & \\ & & -2h \end{pmatrix} \begin{pmatrix} & & \\ & & yz \\ & zy & \end{pmatrix} \begin{pmatrix} & & zy \\ & & \\ yz & & \end{pmatrix} \begin{pmatrix} & yz & \\ zy & & \\ & & \end{pmatrix}$ (O)
$C_2^\rho \parallel \rho$	$T = 23$	yes	A $E^{(1)}$ $E^{(2)}$ $F(x;R_x)$ $F(y=X;R_y=R_x)$ $F(z=x;R_z=R_x)$
$C_2^\rho \parallel \rho$	$T_h = m3$	no	A_g $E_g^{(1)}$ $E_g^{(2)}$ $F_g^{(1)}(R_x)$ $F_g^{(2)}(R_y=R_x)$ $F_g^{(3)}(R_z=R_x)$ $F_u(x=y=z)$
			$\begin{pmatrix} a & & \\ & a & \\ & & a \end{pmatrix} \begin{pmatrix} 2b+d & & \\ & 2b+d & \\ & & -4b-2d \end{pmatrix} \begin{pmatrix} -2\sqrt{3}b-\sqrt{3}d & & \\ & 2\sqrt{3}b+\sqrt{3}d & \\ & & \end{pmatrix} \begin{pmatrix} & & \\ & & f \\ & f & \end{pmatrix} \begin{pmatrix} & & f \\ & & \\ f & & \end{pmatrix} \begin{pmatrix} & f & \\ f & & \\ & & \end{pmatrix} \begin{pmatrix} & & \\ & & e \\ & -e & \end{pmatrix}$
$C_4^\rho \parallel \rho$	$O = 432$	yes	A_1 $E^{(1)}$ $E^{(2)}$ $F_2^{(1)}$ $F_2^{(2)}$ $F_2^{(3)}$ $F_1(x;R_x)$
$S_4^\rho \parallel \rho$	$T_d = \bar{4}3m$	no	A_1 $E^{(1)}$ $E^{(2)}$ $F_2(x)$ $F_2(y=x)$ $F_2(z=x)$ $F_1(R_x)$
$C_4^\rho \parallel \rho$	$O_h = m3m$	no	A_{1g} $E_g^{(1)}$ $E_g^{(2)}$ $F_{2g}^{(1)}$ $F_{2g}^{(2)}$ $F_{2g}^{(3)}$ $F_{1g}(R_x)$
			$\begin{pmatrix} & & -e \\ & & \\ e & & \end{pmatrix} \begin{pmatrix} & e & \\ -e & & \\ & & \end{pmatrix}$ (O)
			$F_1(y=x;R_y=R_x)$ $F_1(z=x;R_z=R_x)$
			$F_1(R_y=R_x)$ $F_1(R_z=R_x)$
			$F_{1g}(R_y=R_x)$ $F_{1g}(R_z=R_x)$ $F_{1u}(x=y=z)$

References

/1/ Ludwig, W.: Festkörperphysik. Frankfurt/Main, Akad. Verl.-Ges., 1970.

/2/ Hopfield, J.J.: Phys. Rev. 112 (1958) 1555.

/3/ Hopfield, J.J.: Proc. of the Int. Conf. on Physics of Semiconductors. Tokyo, 1966.

/4/ Tosatti, E.: Proc. of the Int. School of Physics Enrico Fermi. Course LII, 1971.

/5/ Kittel, Ch.: Quantum Theory of Solids. Sydney-New York-London, J. Wiley & Sons, 1963.

/6/ Lang, M.: Phys. Rev. B2 (1970) 4022.

/7/ Henry, C.H., Hopfield, J.J.: Phys. Rev. Letters 15 (1965) 964.

/8/ Raman, C.V.: Nature (London) 121 (1928) 619.

/9/ Smekal, A.: Naturwiss. 11 (1923) 873.

/10/ Kramers, H.A., Heisenberg, W.: Z. Physik 31 (1925) 681.

/11/ Dirac, P.A.M.: Proc. Roy. Soc. (London) 114 (1927) 710.

/12/ Placzek, G.: in Handb. der Radiologie, ed. E. Marx, Leipzig, Akad. Verl.-Ges., Band VI, Teil II, p. 205, 1934.

/13/ Wilson, E.B. jr., Decius, J.C., Cross, P.C.: Molecular Vibrations. The Theory of Infrared and Raman Vibrational Spectra. London, Mc Graw-Hill Book Comp., Inc. 1955.

/14/ Brandmüller, J., Moser, H.: Einführung in die Ramanspektroskopie. Darmstadt, Steinkopf Verl., 1962.

/15/ Altmann, K., Strey, G., Hochenbleicher, J.G., Brandmüller, J.: Z. Naturforsch. 27a (1972) 56.

/16/ Rasetti, F.: Nature (London) 123 (1929) 205.

/17/ Fast, H., Welsh, H.L., Lepard, D.W.: Can. J. Phys. 47 (1969) 2879.

/18/ Koningstein, J.A., Mortensen, O.S.: in The Raman Effect, ed. Anderson, A., Vol.2, New York, Dekker, 1973.

/19/ Mooradian, A., Wright, G.B.: Phys. Rev. Letters 16 (1966) 999.

/20/ Mooradian, A., Mc Whorter, A.L.: Phys. Rev. Letters 19 (1967) 849.

/21/ Light Scattering Spectra of Solids, ed. G.B. Wright, Berlin-Heidelber-New York, Springer Verlag, 1969.

/22/ Porto, S.P.S., Fleury, P.A., Damen, T.C.: Phys. Rev. 154 (1967) 522.

/23/ Slusher, R.E., Patel, C.K.N., Fleury, P.A.: Phys. Rev. Letters 18 (1967) 77.

/24/ Behringer, J., Brandmüller, J.: Z. Elektrochemie, Ber. der Bunsengesellschaft 60 (1956) 643.

/25/ Behringer, J.: Z. Elektrochemie, Ber. der Bunsengesellschaft 62 (1958) 544; and 'Theorie der molekularen Lichtstreuung', Manuscript, Sektion Physik der Universität, Lehrstuhl J. Brandmüller, Munich, 1967.

/26/ Woodbury, E.J., NG, W.K.: I R E 50 (1962) 2367.

/27/ Eckhardt, G., Hellwarth, R.W., Mc Clung, F.J., Schwarz, S.E., Weiner, D., Woodbury, E.J.: Phys. Rev. Letters 9 (1962) 455.

/28/ Schrötter, H.W.: Naturwiss. 54 (1967) 607.

/29/ Mathieu, J.P.: Spectres de vibration et symétrie des molécules et des cristaux. Paris, Hermann, 1945.

/30/ Matossi, F.: Der Raman Effekt. Braunschweig, Friedr. Vieweg, 1959.

/31/ Loudon, R.: Adv. in Phys. 13 (1964) 423.

/32/ Suschtschinskii, M.M.: Raman Spectra of Molecules and Crystals. New York-London-Jerusalem, Heyden & Son Inc., 1972.

/33/ Poulet, H., Mathieu, J.P.: Spectres de vibration et symétrie des cristaux. Paris-London-New York, Gordon & Breach, 1970.

/34/ Balkanski, M. ed.: Light Scattering in Solids. Flammarion, Paris, 1971.

/35/ Brandmüller, J., Schrötter, H.W.: Fortschr. der chem. Forschung, - Topics in Chemistry 36 (1973) 85.

/36/ Huang, K.: Nature 167 (1951), 779.
Proc. Roy. Soc. A208 (1951) 352.

/37/ Leibfried, G.: in Handb. der Physik, ed. S. Flügge, Berlin-Heidelberg-New York, Springer Verlag, Band VII/1, p.164, 1955.

/38/ Süssmann, G.: Z. Naturforsch. 11a (1956) 1.

/39/ Behringer, J.: Springer Tracts in Modern Physics 68 (1973) 161.

/40/ Int. Tables for X-Ray Crystallography, Vol. I, 3rd Edition, ed. N.F.M. Henry and K. Lonsdale, Birmingham, The Kynoch Press, 1969.

/41/ De Boer, J.L., Bolhuis, F.v., Olthof-Hazekamp, R., Vos, A.: Acta Cryst. 21 (1966) 841.

/42/ Rosenzweig, A., Moresin, B.: Acta Cryst. 20 (1966) 578.

/43/ Matossi, F.: Gruppentheorie der Eigenschwingungen von Punktsystemen. Berlin-Heidelberg-Göttingen, Springer Verlag, 1961.

/44/ Mc Weeny, R., Symmetry. An Introduction to Group Theory and its Applications. Oxford-London-New York-Paris, Pergamon, 1963.

/45/ Nye, J.F.: Physical Properties of Crystals. Oxford, Clarendon Press, 1969.

/46/ Mortensen, O.S., Koningstein, J.A.: J. Chem. Phys. 98 (1961) 3971.

/47/ Behringer, J.: private communication.

/48/ Rosenthal, J.E., Murphy, G.M.: Rev. Mod. Phys. 8 (1936) 317.

/49/ Christie, J.H., Loockwood, D.J.: J. Chem. Phys. 54 (1971) 1141.

/50/ Menzies, A.C.: private communication 1972.

/51/ Menzies, A.C.: J. Chem. Phys. 23 (1955) 1997.

/52/ Tinkham, M.: Group Theory and Quantum Mechanics. New York-San Francisco-Toronto-London, Mc Graw-Hill Book Comp., 1964.

/53/ Albrecht, A.C.: J. Chem. Phys. 34 (1961) 1476.

/54/ Koningstein, J.A.: Chem. Phys. Letters 2 (1968) 31.

/55/ Child, M.S., Longuet-Higgins, H.C.: Phil. Trans. Roy. Soc. (London) A254 (1961) 259.

/56/ Mc Clain, W.M.: J. Chem. Phys. 55 (1971) 2789.

/57/ Winter, F.X., Brandmüller, J.: unpub.

/58/ Ovander, L.N.: Opt. and Spectr. 9 (1960) 302.

/59/ Loudon, R.: Proc. Roy. Soc. (London) A275 (1963) 218.

/60/ Born, M., Kârmân, T.v.: Phys. Z. 13 (1912) 297.

/61/ Born, M.: Optik. Berlin-Heidelberg-New York, Springer Verlag, 1965.

/62/ Born, M., Huang, K.: Dynamical Theory of Crystal Lattices. Oxford, Clarendon Press, 1962.

/63/ Barker, A.S. jr.: Phys. Rev. 136 (1964) 1290.

/64/ Merten, L.: Z. Naturforsch. 15a (1960) 47.

/65/ Merten, L.: Z. Naturforsch. 16a (1961) 447.

/66/ Merten, L.: Z. Naturforsch. 22a (1967) 359.

/67/ Merten, L.: Phys. Stat. Sol. (b) 30 (1968) 449.

/68/ Lamprecht, G., Merten, L.: Phys. Stat. Sol. (b) 35 (1969) 363.

/69/ Borstel, G., Merten, L.: Z. Naturforsch. 26a (1971) 653.

/70/ Merten, L., Borstel, G.: Z. Naturforsch. 27a (1972) 1073.

/71/ Scott, J.F.: Amer. J. Phys. 39 (1971) 1360.

/72/ Borstel, G., Merten, L.: Light Scattering in Solids, ed. M. Balkanski, Paris, Flammarion, p.247, 1971.

/73/ Barker, A.S., Loudon, R.: Rev. Mod. Phys. 44 (1972) 18.

/74/ Merten, L.: Festkörperprobleme XII, ed. O. Madelung, Oxford-Braunschweig, Pergamon-Vieweg, p.343, 1972.

/75/ Claus, R.: Festkörperprobleme XII, ed. O. Madelung, Oxford-Braunschweig, Pergamon-Vieweg, p.381, 1972.

/76/ Claus, R.: Phys. Stat. Sol. (b) 50 (1972) 11.

/77/ Claus, R.: in Proc. of the Int. School of Physics, Enrico Fermi, ed. S. Califano, New York-London, Academic Press, Course LV, 1972.

/78/ Merten, L.: in Proc. of the Int. School of Physics, Enrico Fermi, ed. E. Burstein, New York-London, Academic Press, Course LII, 1972.

/79/ Cochran, W., Cowley, R.A.: J. Phys. Chem. Solids 23 (1962), 447.

/80/ Pick, R.M.: Adv. Phys. 19 (1970) 269.

/81/ Kurosawa, T.: J. Phys. Soc. Japan 16 (1961) 1298.

/82/ Leite,R.C.C.,Damen,T.C.,Scott,J.F.: in Light Scattering Spectra of Solids, ed. G.B.Wright, New York, Springer-Verlag,p.359,1969

/83/ Scott, J.F., Fleury, P.A., Worlock, J.M.: Phys. Rev. 177 (1969) 1288.

/84/ Fleury, P.A., Worlock, J.M.: Phys. Rev. Letters 18 (1967) 665.

/85/ Claus, R.: Rev. Sci. Instr. 42 (1971) 341.

/86/ Claus, R.: Fachberichte der 35. Physikertagung, Hannover 1970. Stuttgart, Teubner, 1970.

/87/ Onstott, J., Lucovsky, G.: J. Phys. Chem. Solids 31 (1970)2755.

/88/ Merten, L., Lamprecht, G.: Phys. Stat. Sol. 39 (1970) 573.

/89/ Poulet, H.: C.R. Acad. Sci. Paris 234 (1952) 2185;
Ann. Phys. Paris 10 (1955) 908.

/90/ Damen, T.C., Porto, S.P.S., Tell, B.: Phys. Rev. 142 (1966)570.

/91/ Scott, J.F., Porto, S.P.S.: Phys. Rev. 161 (1967) 903.

/92/ Merten, L., Lamprecht, G.: Z. Naturforsch. 26a (1971) 215.

/93/ Hartwig, C.M., Wiener-Avnear, E., Smit, J., Porto, S.P.S.: Phys. Rev. B3 (1971) 2078.

/94/ Unger, B., Schaack, G.: Phys. Letters 33A (1970) 295.

/95/ Unger, B.: Phys. Stat. Sol. (b) 49 (1972) 107.

/96/ Otaguro, W., Arguello, C.A., Porto, S.P.S.: Phys. Rev. B1 (1970) 2818.

/97/ Otaguro, W., Wiener-Avnear, E., Arguello, C.A., Porto, S.P.S.: Phys. Rev. B4 (1971) 4542.

/98/ Richter, W.: J. Phys. & Chem. Solids (GB) 33 (1972) 2123.

/99/ Claus, R., Borstel, G., Wiesendanger, E., Steffan, L.: Z. Naturforsch. 27a (1972) 1186.

/100/ Claus, R., Borstel, G., Wiesendanger, E., Steffan, L.: Phys. Rev. B6 (1972) 4878.

/101/ Borstel, G., Merten, L.: in Polaritons, Proceedings of the First Taormina Research Conference on the Structure of Matter, ed. E. Burstein and F. De Martini, New York, Pergamon, 1974.

/102/ Claus, R., Winter, F.X.: in Polaritons, Proceedings of the First Taormina Research Conference on the Structure of Matter, ed. E. Burstein and F. De Martini, New York, Pergamon, 1974.

/103/ Posledovich, M., Winter, F.X., Borstel, G., Claus, R.: Phys. Stat. Sol. (b) 55 (1973) 711.

/104/ Porto, S.P.S., Tell, B., Damen, T.C.: Phys. Rev. Letters 16 (1966) 450.

/105/ Claus, R.: Z. Naturforsch. 25a (1970) 306.

/106/ Scott, J.F., Cheesman, L.E., Porto, S.P.S.: Phys. Rev. 162 (1967) 834.

/107/ Pinczuk, A., Burstein, E., Ushioda, S.: Sol. State Comm. 7 (1969) 139.

/108/ Winter, F.X., Claus, R.: Opt. Comm. 6 (1972) 22.

/109/ Akhamanov, S.A., Fadeev, V.V., Khokhlov, R.V., Chunaev, O.N.: Soviet Phys. - JETP Letters 6 (1967) 85.

/110/ Obukhovskii, V.V., Ponath, H., Strizhevskii, V.L.: Phys. Stat. Sol. (b) 41 (1970) 837.

/111/ Harris, S.E., Oshman, M.K., Byer, R.L.: Phys. Rev. Letters 18 (1967) 732.

/112/ Magde, D., Mahr, H.: Phys. Rev. Letters 18 (1967) 905.

/113/ Byer, R.L., Harris, S.E.: Phys. Rev. 168 (1968) 1064.

/114/ Magde, D., Mahr, H.: Phys. Rev. 171 (1968) 393.

/115/ Giallorenzi, T.G., Tang, C.L.: Appl. Phys. Letters 12 (1968) 376.

/116/ Klyshko, D.N., Krindach, D.P.: Opt. Spectr. 26 (1969) 532.

/117/ Giallorenzi, T.G., Tang, C.L.: Phys. Rev. 184 (1969) 353.

/118/ Klyshko, D.N., Penin, A.N., Polkovnikov, B.F.: Soviet Physics - JETP Letters 11 (1970) 5.

/119/ Winter, F.X.: Diplomathesis, Munich, 1972.

/120/ Claus, R., Borstel, G., Merten, L.: Opt. Comm. 3 (1971) 17.

/121/ Winter, F.X.: Phys. Letters 40A (1972) 425.

/122/ Ponath, H.E., Kneipp, K.D., Kühmstedt, R.: in Laser und ihre Anwendungen, Berlin, Deutsche Akademie der Wissenschaften 1970.

/123/ Ponath, H.E., Kneipp, K.D.: Exp. Techn. Phys., to be published.

/124/ Winter, F.X.: private communication.

/125/ Graf, L., Schaack, G., Unger, B.: Phys. Stat. Sol. (b) 54 (1972) 261.

/126/ Ranson, P., Peretti, P., Rousset, Y., Koningstein, J.A.: Chem. Phys. Letters 16 (1972) 396.

/127/ Balkanski, M., Teng, M.K., Massot, M., Shapiro, S.M.: in Light Scattering in Solids, ed. M. Balkanski, Paris, Flammarion,1971.

/128/ Dobrzhanskii, G.F., Kitaeva, V.F., Kuleskii, L.A., Polivanov, Yu.N., Prokhorov, A.M., Sovolev, N.N.: ZhETF Pis. Red. 12 (1970) 505.

/129/ Krauzman, M.: C. R. Acad. Sci. Paris, Ser. B (1970) 856.

/130/ Couture, L., Krauzman, M., Mathieu, J.P.: C. R. Acad. Sci. Paris, Ser. B (1970) 1246.

/131/ Klyshko, D.N., Kutsov, V.F., Penin, A.N., Polnikov, B.F.: Soviet Phys. - JETP 35 (1970) 960.

/132/ Kitaeva, V.F., Kulevskii, L.A., Polivanov, Yu.N., Poluektov, S.N.: Soviet Phys. - JETP Letters 16 (1972) 383.

/133/ Mavrin, B.N., Sterin, Kh.E.: Soviet Phys. Sol. State 14 (1973) 2402.

/134/ Merten, L.: Z. Naturforsch. 24a (1969) 1878.

/135/ Merten, L.: Phys. Stat. Sol. (b) 28 (1968) 111.

/136/ Merten, L.: Phys. Stat. Sol. (b) 25 (1968) 125.

/137/ Winter, F.X., Wiesendanger, E., Claus, R.: Phys. Stat. Sol. (b) 64 (1974) 95.

/138/ Landau, L.D., Lifschitz, E.M.: Lehrbuch der Theoretischen Physik. Berlin, Akademie Verlag, Bd. VIII, §11, 1971.

/139/ Merten, L., Borstel, G.: Z. Naturforsch. 27a (1972) 1792.

/140/ Bilz, H., Genzel, L., Hopp, H.: Z. Phys. 160 (1960) 535.

/141/ Cowley, R.A.: Adv. Phys. 12 (1963) 427.

/142/ Maradudin, A.A., Weiss, G.H., Jepsen, D.W.: J. Math. Phys. 2 (1961) 349.

/143/ Barker, A.S.: Phys. Rev. 165 (1968) 917.

/144/ Coffinet, I.P., De Martini, F.: Phys. Rev. Letters 20 (1969)60.

/145/ Ushioda, S., Mc Mullen, J.D.: Sol. State Comm. 11 (1972) 299.

/146/ Lamprecht, G., Merten, L.: Phys. Stat. Sol. (b) 55 (1973) 33.

/147/ Merten, L.: Phys. Stat. Sol. (b) 55 (1973) K 143.

/148/ Alfano, R.R.: J. Opt. Soc. Amer. 60 (1970) 66.

/149/ Alfano, R.R., Giallorenzi, T.G.: Opt. Comm. 4 (1971) 271.

/150/ Benson, H.J., Mills, D.L.: Phys. Rev. B1 (1970) 950.

/151/ Andrade, P.Da.R., Prasad Rao, A.D., Katiyar, R.S., Porto, S. P.S.: Sol. State Comm. 12 (1973) 847.

/152/ Andrade, P.Da.R., Porto, S.P.S.: Sol.State Comm. 13 (1973) 1249.

/153/ Merten, L., Andrade, P.Da.R.: Phys.Stat.Sol.(b) 62 (1974) 283.

/154/ Borstel, G., Andrade, P.Da.R., Merten, L.: Phys. Stat. Sol. (b) 71 (1975) in press.

/155/ Behringer, J.: J. Raman spectr. 2 (1974) 275.

/156/ Merten, L., Andrade, P.Da.R., Borstel, G.: Phys. Stat. Sol. (b) 70 (1975) in press.

/157/ Chaves, A.S., Porto, S.P.S.: Sol. State Comm. 13 (1973) 865.

/158/ Unger, B., Schaack, G.: Phys. Stat. Sol. (b) 48 (1971) 285.

/159/ Borstel, G., Merten, L.: Z. Naturforsch. 28a (1973) 1038.

/160/ Loudon, R.: Light Scattering Spectra of Solids, ed. G.B. Wright, New York, Springer Verlag, p.25, 1969.

/161/ Borstel, G.: Z. Naturforsch. 28a (1973) 1055.

/162/ Loudon, R.: Proc. Phys. Soc. (London) 82 (1963) 393.

/163/ Henry, C.H., Garrett, C.G.P.: Phys. Rev. 171 (1968) 1058.

/164/ Claus, R.: to be published.

/165/ Ovander, L.N.: Usp. Fiz. Nauk 86 (1965) 3.

/166/ Burstein, E., Ushioda, S., Pinczuk, A.: Light Scattering Spectra of Solids, ed. G.B. Wright, New York, Springer Verlag, 1969.

/167/ Bloembergen, N., Shen, Y.R.: Phys. Rev. Letters 12 (1964) 504.

/168/ Bloembergen, N.: Nonlinear Optics. New York-Amsterdam, Benjamin, 1965.

/169/ Zeiger, H.J., Tannenwald, P.E., Kern, S., Heredeen, R.: Phys. Rev. Letters 11 (1963) 419.

/170/ Chiao, R.Y., Townes, C.H.: Phys. Rev. Letters 12 (1964) 592.

/171/ Chiao, R.Y., Stoicheff, B.P.: Phys. Rev. Letters 12 (1964)290.

/172/ Ataev, B.M., Lugovoi, V.N.: Opt. and Spectr. 27 (1969) 380.

/173/ Schrötter, H.W.: Naturwissenschaften 19 (1967) 513.

/174/ Rath, K.: Diplomathesis, Münster 1973.

/175/ Bloembergen, N.: Am. J. Phys. 35 (1967) 989.

/176/ Faust, W.L., Henry, C.H., Eich, R.H.: Phys. Rev. 173 (1968) 781.

/177/ Kurtz, S.K., Giordmaine, J.A.: Phys. Rev. Letters 22 (1969) 192.

/178/ Yarborough, J.M., Sussman, S.S., Puthoff, H.E., Pantell, R.H., Johnson, B.C.: Appl. Phys. Letters 15 (1969) 102.

/179/ Gelbwachs, J., Pantell, R.H., Puthoff, H.E., Yarborough, J.M.: Appl. Phys. Letters 14 (1969) 258.

/180/ Johnson, B.C., Puthoff, H.E., Soottoo, J., Sussman, S.S.: Appl. Phys. Letters 18 (1971) 181.

/181/ Schrötter, H.W.: Z. Naturforsch. 26a (1971) 165.

/182/ Amzallag, E., Chang, T.S., Johnson, B.C., Pantell, R.H., Puthoff, H.E.: J. Appl. Phys. 42 (1971) 3251.

/183/ De Martini, F.: Phys. Letters 30A (1969) 319.

/184/ De Martini, F.: Phys. Letters 30A (1969) 547.

/185/ De Martini, F.: Phys. Rev. B4 (1971) 4556.

/186/ Hwang, D.M., Solin, S.A.: Phys. Rev. B9 (1974) 1884.

/187/ Lauberau, A., von der Linde, D., Kaiser, W.: Opt. Comm. 7 (1973) 173.

/188/ Karmenyan,K.V., Chilingaryan, Yu.S.: Soviet Phys. - JETP Letters 17 (1973) 73.

/189/ Varga, B.B.: Phys. Rev. 137 (1965) 1896.

/190/ Patel, C.K.N., Slusher, R.E.: Phys. Rev. Letters 22 (1969) 282.

/191/ Shah, J., Damen, T.C., Scott, J.F., Leite, R.C.C.: Phys. Rev. B3 (1971) 4238.

/192/ Hermann, J., Kneipp, K.D., Ponath, H.E., Werncke, W., Klein, J., Lau, A.: Exp. Techn. Phys. 22 (1974) 97.

/193/ Falge, H.J., Otto, A., Sohler, W.: Phys. Stat. Sol. (b) 63 (1974) 259.

/194/ Nitsch, W., Falge, H.J., Claus, R.: Z. Naturforsch. 29a (1974) 1011.

/195/ Drude, P.: Ann. Phys. Chem. (N.F.) 32 (1887) 619.

/196/ Mosteller, L.P. jr., Wooten, F.: J. Opt. Soc. Amer. 58 (1968) 511.

/197/ Balge, H.J.: Thesis, Munich 1974.

/198/ Marschall, N., Fischer, B.: Phys. Rev. Letters 28 (1972) 811.

/199/ Bryksin, V.V., Gerbstein, Yu.M., Mirlin, D.N.: Phys. Stat. Sol. (b) 51 (1972) 901.

/200/ Nitsch, W., Claus, R.: Z. Naturforsch. 29a (1974) 1017.

/201/ Mills, D.L., Maradudin, A.A.: Phys. Rev. B1 (1970) 903.

/202/ Ohtaka, K.: Phys. Stat. Sol. (b) 57 (1973) 51.

/203/ Hwang, D.M.: Phys. Rev. B9 (1974) 2717.

/204/ Agranovich, V.M., Lalov, I.I.: Sov. Phys. Sol. State 13 (1971) 859.

/205/ Kitaeva, V.F., Kulevskii, L.A., Polivanov, Yu.N., Poluektov, S.N.: Soviet Physics - JETP Letters 16 (1972) 15.

/206/ Mavrin, B.N., Sterin, Kh.E.: Soviet Physics - JETP Letters 16 (1973) 50.

/207/ Kneipp, K.D., Ponath, H.E., Strizhevskii, V.L., Yashkir, Yu.N.: Soviet Physics - JETP Letters 18 (1973) 50.

/208/ Chang, T.S., Johnson, B.C., Amzallag, E., Pantell, R.H., Rokni, M., Wall, L.S.: Opt. Comm. 4 (1971) 72.

/209/ Rokni, M., Wall, C.S., Amzallag, E., Chang, T.S.: Sol. State Comm. 10 (1972) 103.

/210/ Laughman, L., Davis, L.W., Nakamura, T.: Phys. Rev. B6 (1972) 3322.

/211/ Burns, G.: Phys. Letters 43A (1973) 271.

/212/ Heiman, D., Ushioda, S.: Phys. Rev. B9 (1974) 2122.

/213/ Faust, W.L., Henry, C.H.: Phys. Rev. Letters 17 (1966) 1265.

/214/ Puthoff, H.E., Pantell, R.H., Huth, B.G., Chacon, M.A.: J. Appl. Phys. 39 (1968) 2144.

/215/ Scott, J.F., Ushioda, S.: Light Scattering Spectra of Solids, ed. G.B. Wright, New York, Springer Verlag, 1969.

/216/ Tsuboi, M., Terada, M., Kajiura, T.: Bull. Chem. Soc. Japan 42 (1969) 1871.

/217/ Ohtaka, K., Fujiwara, T.: J. Phys. Soc. Japan 27 (1969) 901.

/218/ Ohama, N., Okamoto, Y.: J. Phys. Soc. Japan 29 (1970) 1648.

/219/ Benson, H.J., Mills, D.L.: Sol. State Comm. 8 (1970) 1387.

/220/ Bendow, B., Birman, J.L.: Phys. Rev. B1 (1970) 1678.

/221/ Bendow, B.: Phys. Rev. B2 (1970) 5051.

/222/ Khashkhozhev, Z.M., Lemanov, V.V., Pisarev, R.V.: Soviet Phys. Sol. State 12 (1970) 941.

/223/ Asawa, C.K.: Phys. Rev. B2 (1970) 2068.

/224/ Asawa, C.K., Barnoski, M.K.: Phys. Rev. B3 (1971) 2682.

/225/ Otaguro, W.S., Wiener-Avnear, E., Porto, S.P.S.: Appl. Phys. Letters 18 (1971) 499.

/226/ Shepherd, I.W.: Sol. State Comm. 9 (1971) 1857.

/227/ Anda, E.: Sol. State Comm. 9 (1971) 1545.

/228/ Barentzen, H., Schrader, B., Merten, L.: Phys. Stat. Sol. (b) 45 (1971) 505.

/229/ Tsu, R., Iha, S.S.: Appl. Phys. Letters 20 (1972) 16.

/230/ Giallorenzi, T.G.: Phys. Rev. B5 (1972) 2314.

/231/ Reinisch, R., Paraire, N., Biraud-Laval, S.: C. R. Acad. Sci. Paris, Ser. B, t 275 (1972) 829.

/232/ Burns, G.: Appl. Phys. Letters 20 (1972) 230.

/233/ Mavrin, B.N., Abramovich, T.E., Sterin, Kh.E.: Soviet Phys. Sol. State 14 (1972) 1562.

/234/ D'Andrea, A., Fornari, B., Mattei, G., Pagannone, M., Scrocco, M.: Phys. Stat. Sol. (b) 54 (1972) K 131.

/235/ Agranovich, V.M., Ginzburg, V.L.: Soviet Physics - JETP 34 (1972) 662.

/236/ Graf, L., Schaack, G., Unger, B.: Adv. in Raman Spectroscopy. London, Heyden & Son, Vol. I, 1972.

/237/ Krauzman, M., Le Postollec, M., Mathieu, J.P.: Phys. Stat. Sol. (b) 60 (1973) 761.

/238/ Hwang, D.M., Solin, S.A.: Sol. State Comm. 13 (1973) 983.

/239/ Ushioda, S., Mc Mullen, J.D., Delaney, M.J.: Phys. Rev. B8 (1973) 4634.

/240/ Sindeev, Yu.G.: Soviet Phys. Sol. State 14 (1973) 3110.

/241/ Strizhevskii, V.L., Yashkir, Yu.N.: Phys. Stat. Sol. (b) 61 (1974) 353.

/242/ Otaguro, W.S., Wiener-Avnear, E., Porto, S.P.S., Smit, J.: Phys. Rev. B6 (1972) 3100.

/243/ Inoue, M.: Phys. Letters 47A (1974) 311.

/244/ Schopper, H.: Z. Phys. 132 (1952) 146.

/245/ Flournoy, P.A., Schaffers, W.J.: Spectrochim. Acta 22 (1966) 5.

/246/ Mosteller, L.P. jr., Wooten, F.: J. Opt. Soc. Amer. 58 (1968) 511.

/247/ Decius, J.C., Frech, R., Brüesch, P.: J. Chem. Phys. 58 (1973) 4056.

/248/ Sohler, W.: Opt. Comm. 10 (1974) 203.

/249/ Szivessy, G.: Handbuch der Physik, ed. S. Flügge, Berlin, Springer Verlag, Vol. 20, Kap. 4, 1928.

/250/ Wolter, H.: Handbuch der Physik, ed. S. Flügge, Berlin, Springer Verlag, Vol. 24, p.555, 1956.

/251/ Bell, E.E.: in Handbuch der Physik, ed. S. Flügge, Berlin, Springer Verlag, Vol. 25, 2a, p.1, 1967.

/252/ Koch, E.E., Otto, A., Kliewer, K.L.: Chem. Phys. 3 (1974) 362.

/253/ Borstel, G., Falge, H.J., Otto, A.: Springer Tracts in Modern Physics 74 (1974) 107.

/254/ Hizhnyakov, V.V.: Phys. Stat. Sol. (b) 34 (1969) 421.

/255/ Mavroyannis, C.: Phys. Rev. B1 (1970) 3439.

/256/ Inomata, H., Horie, C.: Phys. Letters 31A (1970) 418.

/257/ Stern, E.A., Ferrell, R.A.: Phys. Rev. 120 (1960) 130.

/258/ Falge, H.J., Otto,A.: Phys. Stat. Sol. (b) 56 (1973) 523.

/259/ Borstel, G.: Thesis, Münster, 1974.

/260/ Otto, A.: Optik 38 (1973) 566.

/261/ Lyubimov, V.N., Samikov, D.G.: Soviet Phys. Solid State 14 (1972) 574.

/262/ Hartstein, A., Burstein, E., Brion, J.J., Wallis, R.F.: in Polaritons, Proc. of the First Taormina Conf. on the Structure of Matter, ed. E. Burstein and F. De Martini, New York, Pergamon, p.111, 1974.

/263/ Kellermann, E.W.: Phil. Trans. Roy. Soc. 238 (1940) 513;
Proc. Roy. Soc. A178 (1941) 17.

/264/ Dick, B. jr., Overhauser, A.: Phys. Rev. 112 (1958) 90.

/265/ Cochran, W.: Proc. Roy. Soc. A253 (1959) 260.

/266/ Cochran, W.: Adv. in Phys. 9 (1960) 387.

/267/ Cochran, W.: Adv. in Phys. 10 (1961) 401.

/268/ Schröder, U.: Sol. State Comm. 4 (1966) 347.

/269/ Nüsslein, V., Schröder, U.: Phys. Stat. Sol. 21 (1967) 309.

/270/ Phillips, J.C.: Phys. Rev. 166 (1968) 832.

/271/ Phillips, J.C.: Phys. Rev. 168 (1968) 917.

/272/ Martin, R.M.: Phys. Rev. Letters 21 (1968) 536.

/273/ Martin, R.M.: Phys. Rev. 186 (1969) 871.

/274/ Martin, R.M.: Phys. Rev. B1 (1970) 910.

/275/ Weber, W.: Phys. Rev. Letters 33 (1974) 371.

/276/ Sham, L.J.: Phys. Rev. 188 (1969) 1931.

/277/ Pick, R.M., Cohen, M.H., Martin, R.M.: Phys. Rev. B1 (1970) 910.

/278/ Jones, W.J., Stoicheff, B.P.: Phys. Rev. Letters 13 (1964) 657.

/279/ Dumartin, S., Oksengorn, B., Vodar, B.: C. R. Acad. Sci. Paris 261 (1967) 3767.

/280/ Long, D., Stanton, L.: Proc. Roy. Soc. A318 (1970) 441.

/281/ Scott, J.F., Damen, T.C., Shah, J.: Opt. Comm. 3 (1971) 384.

/282/ Etchepare, J., Matthieu Merian: C. R. Acad. Sci. Paris, Ser.B, t 278 (1974) 1071.

/283/ Biraud-Laval, S.: Le Journal de Physique t.35 (1974) 513.

/284/ Claus, R., Schrötter, H.W.: Light Scattering in Solids, ed. M. Balkanski, Paris, Flammarion, 1971.

/285/ Porto, S.P.S., Giordmaine, J.A., Damen, T.C.: Phys. Rev. 147 (1966) 608.

/286/ Anderson, A.: The Raman Effect, New York, Marcel Dekker, Inc. Vol.1 (1971), Vol.2 (1973).

/287/ Agranovich, V.M.: Soviet Phys.-JETP 19 (1974) 16.

/288/ Tannenwald, P.E., Weinberg, D.L.: J. Q. Elec. 3 (1967) 334.

/289/ Scott, J.F.: J. Q. Elec. 3 (1967) 694.

/290/ Otto, A.: Festkörperprobleme XIV. Advances in Solid State Physics, ed. H.J. Queisser, Oxford-Braunschweig, Pergamon-Vieweg, 1974.

/291/ Scott, J.F.: Rev. Mod. Phys. 46 (1974) 83.

/292/ Walker, Ch.T., Slack, G.A.: Amer. J. Phys. 38 (1970) 1380.

/293/ Maradudin, A.A., Fine, A.E.: Phys. Rev. 128 (1962) 2589.

/294/ Maradudin, A.A., Fine, A.E., Vineyard, G.H.: Phys. Stat. Sol. 2 (1962) 1479.

/295/ Maradudin, A.A., Mills, D.L.: Phys. Rev. B7 (1973) 2787.

/296/ Maradudin, A.A., Wallis, R.F.: Phys. Rev. 125 (1962) 1277.

/297/ Cowley, R.A.: in Lattice Dynamics, ed. R.F. Wallis, New York, Pergamon Press, 1963.

/298/ Petersson, J.: Proc. Europ. Meeting on Ferroelectrics, Saarbrücken, 1969. Stuttgart, Wiss. Verl. Handl., p.84, 1970.

/299/ Petersson, J.: Ferroelectrics 4 (1972) 221.

/300/ Mills, D.L., Burstein, E.: Rep. Prog. Phys. 37 (1974) 817.

/301/ Ushioda, S., Pinczuk, A., Burstein, E., Mills, D.L.: in Light Scattering Spectra of Solids, ed. G.B. Wright, New York, Springer Verlag, p. 347, 1969.

/302/ Fano, U.: Phys. Rev. 103 (1956) 1202.

/303/ Hopfield, J.J.: Proc. Int. School of Physics, Enrico Fermi, New York-London, Academic Press, Course XLII, p.340, 1969.

/304/ Landau, L.D., Lifschitz, E.M.: Lehrbuch der Theor. Phys., Berlin, Akademie Verlag, Bd.V, §11, 1971.

/305/ Behringer, J.: in Molecular Spectroscopy, Vol.II, ed. R.F. Barrow, D.A. Long, D.J. Millen, The Chemical Society, London 1974.

/306/ Begley, R.F., Byer, R.L., Harvey, A.B.: 4th Int. Conf. on Raman Spectroscopy, Brunswick, Main, USA, 1974.

/307/ Nyquist, H.: Phys. Rev. 32 (1928) 110.

/308/ De Martini, F.: in Proc. of the Int. School of Physics 'Enrico Fermi', Course LII, ed. E. Burstein, Academic Press, 1972.

/309/ Wallis, R.F.: in Proc. of the Int. School of Physics 'Enrico Fermi', Course LII, ed. E. Burstein, Academic Press, 1972.

/310/ Pick, R.: in Proc. of the Int. School of Physics 'Enrico Fermi', Course LII, ed. E. Burstein, Academic Press, 1972.

/311/ Lax, M., Nelson, D.F.: in Proc. of the Int. School of Physics 'Enrico Fermi', Course LII, ed. E. Burstein, Academic Press, 1972.

/312/ Wemple, H.: in Proc. of the Int. School of Physics 'Enrico Fermi', Course LII, ed. E. Burstein, Academic Press, 1972.

/313/ Burstein, E.: in Proc. of the Int. School of Physics 'Enrico Fermi', Course LII, ed. E. Burstein, Academic Press, 1972.

/314/ Loudon, R.: J. Phys. A: Gen. Phys. 3 (1970) 233.

/315/ Porto, S.P.S.: in Light Scattering Spectra of Solids, ed. G.B. Wright, New York, Springer Verlag, 1969.

/316/ Polaritons, Proceedings of the First Taormina Research Conference on the Structure of Matter, ed. E. Burstein and F. De Martini, New York, Pergamon Press, 1974.

/317/ Winter, F.X.: Thesis, Munich, 1975.

/318/ Nitsch, W.: Z. Naturforsch. 30a (1975) 537.

/319/ Birman, J.L.: Sol. State Comm. 13 (1973) 1189.

/320/ Birman, J.L., Benson, Rh.: Phys. Rev. B9 (1974) 4512.

/321/ Birman, J.L.: Phys. Rev. B9 (1974) 4518.

/322/ Birman, J.L.: in Handbuch der Physik, Vol. XXV/2b,Light and Matter lb, ed. L. Genzel, New York, Springer Verlag, 1974.

/323/ Fries, J., Claus, R.: J. Raman Spectr. 1 (1973) 71.

/324/ Madelung, O.: Festkörpertheorie II, Heidelberger Taschenbücher, Berlin-Heidelberg-New York, Springer Verlag, 1972.

/325/ Cochran, W., Cowley, R.A.: in Handbuch der Physik XXV/2a, Light and Matter la, ed. L. Genzel, Springer Verlag, p.59,1967.

/326/ Harris, E.G.: A pedestrian Approach to Quantum Field Theory. New York-Syndney-Toronto, Wiley-Interscience, 1972.

/327/ Heitler, W.: The Quantum Theory of Radiation. Oxford, Clarendon Press, 1954.

/328/ Ewald, P.P.: Ann. Phys. 54 (1917) 519, 557 and 64 (1921) 253; Nachr. Ges. Wiss., Göttingen 55 (1938).

/329/ Laughman, L., Davis, L.W.: Phys. Rev. B10 (1974) 2590.

/330/ Gadow, P., Lau, A., Thuy, Ch.T., Weigmann, H.-J., Werncke, W., Lenz, K., Pfeiffer, M.: Opt. Comm. 4 (1971) 226.

/331/ Louisell, W.: Radiation and Noise in Quantum Electronics. New York-San Francisco-Toronto-London, Mc Graw-Hill, 1964.

/332/ Pekar, S.I.: Soviet Physics - JETP 11 (1960) 1286.

/333/ Koningstein, J.A.: Introduction to the Theory of the Raman Effect. Dordrecht, Holland, Reindl Publ. Comp., 1972.

/334/ Poppinger, M.: Magnonen, Phononen und Excitonen von MnF_2. Thesis, Munich, 1974.

/335/ Szymanski, H.A.: Raman Spectroscopy, Theory and Practice, New York-London, Plenum Press, Vol.1 (1967), Vol.2 (1970).

/336/ Woodward, L.A.: Introduction to the Theory of Molecular Vibrations and Vibrational Spectroscopy. Oxford, Clarendon Press, 1972.

/337/ Kiefer, W.: Appl. Spectroscopy 28 (1974) 115.

/338/ Behringer, J.: in Molecular Spectroscopy, Vol.III, ed. R.F. Barrow, D.A. Long, D.J. Millen, London, The Chemical Society (in press).

/339/ Zak, J., Casher, A., Glück, M., Gur, Y.: The Irreducible Representation of Space Groups. New York, W.A. Benjamin, 1969.

/340/ Behringer, J.: Raman Spektren von Kristallen, Vorlesung München WS 1971/72.

/341/ Miller, S.C., Love, W.F.: Tables of Irreducible Representations of Magnetic Space Groups. Boulder, Colorado, Praett Press, 1967.

/342/ Bhagavantam, S.: Crystal Symmetry and Physical Properties. London-New York, Academic Press, 1966.

/343/ Koster, G.F., Dimmock, J.O., Wheeler, R.B., Statz, H.: Properties of the 32 Point Groups. Cambridge, Mass. 1963.

/344/ Paul, H.: Nichtlineare Optik, Vol. I, II. Berlin, Akademie Verlag, 1973.

/345/ Müser, H.E., Petersson, Y.: Fortschritte der Physik 19 (1971) 559.

/346/ Belikova, G.S., Kulevsky, L., Polivanov, Yu.N., Poluektov, S.N., Prokhorov, K.A., Shigoryn, V.D., Shipulo, G.P.: J. Raman Spectr. 2 (1974) 493.

/347/ Landsberg, G., Mandelstam, L.: Naturwissenschaften 16 (1928) 557, 772.

/348/ Shapiro, S.M., Axe J.D.: Phys. Rev. B6 (1972) 2420.

/349/ Hollis, R.L., Ryan, J.F., Scott, J.F.: Phys. Rev. Letters 34 (1975) 209.

/350/ Claus, R., Winter, F.X.: Ber. der Bunsengesellschaft, in print and Proc. of the 3rd Int. Conf. on Light Scattering in Solids, ed. M. Balkanski, R.C.C. Leite, S.P.S. Porto, Paris, Flammarion 1975.

/351/ Claus, R., Nitsch, W.: Proc. of the 3rd Int. Conf. on Light Scattering in Solids, ed. M. Balkanski, R.C.C. Leite, S.P.S. Porto, Paris, Flammarion 1975.

/352/ Ahrens, K.H.F., Schaack, G., Unger, B.: J. Phys. E, in press.

/353/ Kneipp, K., Werncke, W., Ponath, H.E., Klein, J., Lau, A., Chu Dinh Thuy, Phys. Stat. Sol. (b) 64 (1974) 589.

/354/ Haken, H.: Quantenfeldtheorie des Festkörpers, Stuttgart, Teubner 1973.

/355/ Cochran, W.: CRC Critical Reviews in Solid State Sciences, Vol. 2 (1971) 1.

/356/ Sinha, S.K.: CRC Critical Reviews in Solid State Sciences, Vol. 3 (1973) 273.

/357/ Phillips, J.C.: Rev. Mod. Phys. 42 (1970) 317.

Author Index

Subject Index